Frontispiece:

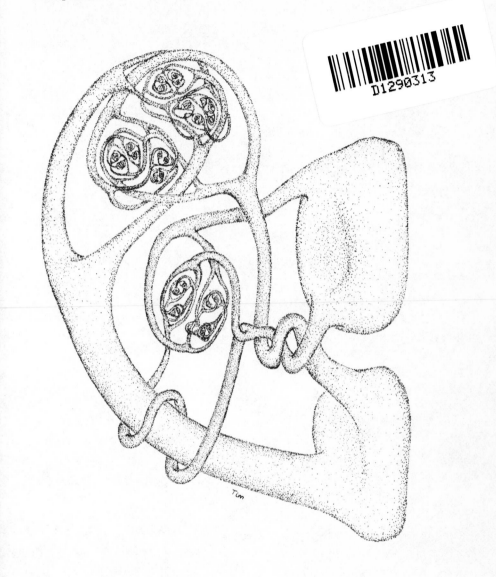

THE MATING DANCE OF THE ALEXANDER HORNED SPHERES   *by* Tim Poston

Seven years of
manifold
1968 - 1980

Seven years of
manifold
1968 - 1980

edited by

IAN STEWART, University of Warwick
JOHN JAWORSKI, BBC Open University Production Centre

Shiva Publishing Limited

SHIVA PUBLISHING LIMITED
4 Church Lane, Nantwich, Cheshire CW5 5RQ, England

IMPRINT EDITIONS
Suite 1104, 415 South Howes St, Fort Collins, CO 80521, USA

British Library Cataloguing in Publication Data

The Best of Manifold.
    1. Mathematics — Periodicals — Addresses, essays, lectures
    I. Stewart, I. N.   II. Jaworski, John
    III. Manifold
    510   QA7

    ISBN 0-906812-07-0

Printed by Devon Print Group, Exeter, Devon

contents

On the Road to MANIFOLD Delay   1

SIGNS   6

Bourbaki   Elizabeth Campbell   7

A new Introduction to Maze Theory   Steven Everett   10

The Annulus Conjecture   Barbara Sands   13

Genuine Fakes   Barry Pilton   18

Green Grow the Fibres-O!   Stan Kelly   20

The Good-natured Rabbit-breeder   Jozef Plojhar   21

Think 13   26

Au Courant with Differential Equations   Tim Poston   27

My bath-tap is a cusp catastrophe   Henry Squarepoint   31

The Fields Medal   Daniel Drolet   32

Sanctification and the Hopf Bifurcation   Philip Holmes   34

15 new ways to catch a Lion   John Barrington   36

A small world   Barry Pilton   40

Portrait of René Thom   Theodor Bröcker   43

The one-move mate   Stanley Collings   44

Hint 1   45

Ontology revisited   Vox Fisher   45

Simple Groups   Eve laChyl   46

Lemmawocky   Carol Lewis   49

Slythaeia Tova   Hassard Dodgson   51

Topology in the Scientist's Toolkit   Christopher Zeeman   52

Hint 2   58

A Pandora's Box of non-Games   Anatole Beck, David Fowler   59

The MANIFOLD guide to Handwaving   62

Meanwhile, back in the Labyrinth...   Steven Everett   64

The 1-colour Theorem   Jozef Plojhar   66

The 2-colour Theorem   Vivienne Hathaway   66

The $[(7+\sqrt{1+48p})/2]$-colour Theorem   68

The 3-colour Theorem   69

The 5-colour Theorem   69

The 4-colour Theorem   Douglas Woodall   69

Another Formula not Representing Primes   Matthew Pordage   76

Come back McGonagall, all is forgiven...   80

An Odd Evening	Ian Stewart	81
Knit Yourself a Klein Bottle	Janis Wanstall	84
Hint 3		84
Beyond the Bounds of Possibility	Michael Forrester	85
Occam's Barber	Roger Hayward	89
The Publication System: a Jaundiced View		90
Think 13: Solutions		91
One-move mate: solution	Stanley Collings	93
Ruler and Compass Construction	Jerry Cornelius	94

2294

"Begin with the tale of MANIFOLD;
The tail is the natural start.
But the Devil whoops, as he whooped of old:
'It's clever, but is it Art?'"
    Sorry, Rudyard.   But we couldn't help it:
We've spent so much of the past decade...

on the Road to MANIFOLD Delay

At some date in 1968, a fledgling mathematical publication nestling
between bright orange covers came into being.   That publication
was the very first MANIFOLD.   The editors had other things on their minds, and no one seems to have recorded the exact date.   A
competition in that issue announces its closing date as April 1st,
hurriedly amended in the editorial to August 1st: Summer 1968 is
possibly the best date that historians will have to work with.

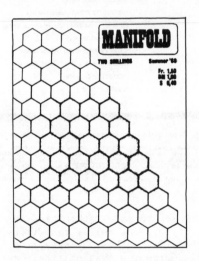

But if the date is not recorded, the place is firmly fixed: the
back kitchen of a large house on Gibbet Hill, on the outskirts of
Coventry.   At that time it housed the Mathematics Institute, University of Warwick.   The Institute, and especially Christopher Zeeman, were very much behind the venture.   They couldn't let us have
any money, but... we inferred that they would not object if we used
duplicator and paper from the Institute's stocks.   Zeeman envisaged
a modest 10-page newssheet: you should have seen his face when we
showed him the first issue - all 56 pages of it!   We printed 200,
which sold at 2/- (now 10p.) each.   They sold so fast that we reprinted a further 150, at 1 a.m. in the kitchen where the duplicat-
or lived.   Zeeman saw the lights burning and came to investigate:
he walked in, assessed the situation, and walked out - all without

a single word of recognition of the event, then or since.

There was no real shortage of mathematical journals, either in 1968 or in 1980 when MANIFOLD (now costing 40p. - well below inflation) finally and gracefully ceased publication. For the main part, these were research documents or undergraduate mathematical society newsletters. MANIFOLD always defied this sort of classification: it has always been - well, MANIFOLD!

MANIFOLD 1

MATHEMATICS INSTITUTE
UNIVERSITY OF WARWICK
COVENTRY

A Mathematics magazine printed and published at the University of Warwick.

MANIFOLD's SIGNS

SMALE & THE VIETNAM WAR: Ramesh Kapadia

THINK OF A NUMBER : Peter Wise

N. BOURBAKI: Elizabeth Campbell

KNIGHT'S TOURS: Jonathan Britt & Jane Mary Grimes

GRUPPEN: "Cosgrove"

INTEGRAL SOLUTIONS: Simon Gazelle

PICTURES IN MATHEMATICS: Brunswick Maths. Inst.

HEX: Michael Jackson

A NOTE ON RECTANGLES: Jeremy Harris

PHOEBUS - What a Name! Josef Plojhar

CRITIQUE by Cosgrove

AUTHORS & STAFF

MANIFOLD is published three times yearly in the Autumn, Spring and Summer.

Price per copy, post free in the United Kingdom, as on front cover. For details of reprints see MANIFOLD's SIGNS. This issue, subject to principle stated within is copyright. SUBSCRIPTION RATES (U.K. only): 10/- for at least three issues, and thereafter until credit runs out. Air-mail subscriptions to USA $3. Further details on all matters to Mathematics Institute, University of Warwick, Coventry, England. Mention "MANIFOLD" in all correspondence.

The aim of MANIFOLD was to make mathematics *accessible*: what made it unique was its style - the belief that it is possible to be *serious* about mathematics without being *solemn*. Its closest model was probably Martin Gardner's 'Mathematical Games' column in *Scientific American*, although MANIFOLD added to this a certain rough-hewn production and a racy, cavalier style of journalism. It used the broad-brush approach: everything from knitting patterns for a Klein Bottle to the Annulus Conjecture. The articles ranged from surveys of research mathematics to the long-running 'Labyrinth' series in which Theseus and the Minotaur debunked some of mathematics's most cherished myths. It had crosswords and cartoons; letters; and MANIFOLD's SIGNS (a feeble pun) specializing in anecdotes, typographical errors, and the occasional piece of lateral thinking, like:

☐ Why isn't the plural 'prooves', like 'hooves' and 'rooves'?
☐ Has it occurred to you that *Concorde* is actually longer than the original Wright brothers' flight?

2

MANIFOLD had a number of strengths, besides the consistently high standard of contribution: two are worth recording. First, it began life in the fertile environment of the Warwick Mathematics Institute. One of the founders remarked that "the environment at Warwick (in 1968) was so rich that you felt that anything you planted would grow". During its 12-year life, MANIFOLD drew on that richness and, until it ceased publication, remained close to the roots of Warwick life.

Its second strength was also its weakness. Most journals of the 'undergraduate mathematics society' genre are doubly handicapped in that one function is to report the doings of a body that varies in health and vitality from year to year, and that their editors are drawn from that same body. MANIFOLD was exclusively a magazine. It had, from the start, a strong editorial board which did not pass on after one year, but stayed with the magazine, nurturing its growth and encouraging the development of its style. Mostly, they were still around in 1980 as the final issue went to press. And that, in the final analysis, was why it closed: the editors had all built careers for themselves and had less time for the buccaneering spirit that had built MANIFOLD.

We were often asked: "where does all this invention come from?" It is time to reveal the answer: *we stole.*

We even stole the title. The *New Statesman* referred to a poetry magazine of the same name in a competition, and we realised that it was just what we needed. By the time the competition's results were published, we had conveniently forgotten the theft and felt suitably indignant about the poetry magazine stealing *our* title.

We stole the cartoonist's name, Cosgrove, from a calendar advertising a local garage, hanging in the duplicator room. Such techniques later became standard: we called them 'taking inspiration from the environment'.

We stole the surname of one Balkan politician and the forename of another to get Jozef Plojhar, a regular but pseudonymous contributor. In fact, we manufactured names wholesale. It seemed self-evident that an entire issue written by one person was a bad thing - so we invented people. We stole Claude Chevalley's surname and created Eve laChyl, our resident algebraist. We stole Barry Pilton, which could have caused problems when he went on to become successful as scriptwriter for the BBC's *Week Ending* satire show, but we thought that that was probably a natural progression. We invented people so successfully that when Ramesh Kapadia joined the editors, most readers spent a profitless ten minutes trying to work out who he was an anagram of.

We stole the *Guardian*'s 'Miscellany' column, calling it MANIFOLD's SIGNS; we stole the style of our competitions from *Punch*; we stole the idea of having in each issue an independent (oh yes?) and unbiased criticism of the articles, from *She* (we dropped the idea almost immediately); we stole jokes:

Q What's purple and commutes?

3

Ⓐ An abelian grape.

But we didn't steal everything: our cartoon strip 'Gruppen', for example - that was ours, and there's a story behind that too.   In 1968, the Mathematical Institute at Oxford had a large empty glass case.   The standing joke was that it was built to house the next simple group discovered.   Not to be outdone, some Warwick graduate students invented and *built* a trap for simple groups.   It was made out of an old shoe-box and populated with little cardboard figures with the names of simple groups written on them.   The Higman-Sims group looked like the beastie on the right, we recall: it's a pity that the Fischer-Griess Monster had not been disco-vered then: the mind boggles! Anyway, Cosgrove based his cartoon characters on

the Higman-Sims
simple group.

these simple groups.   'Groups' sounding too bland, we did as many mathematicians before us, changed to German, and plumped for 'Gruppen'.   Cosgrove still lives, despite erroneous reports of his death in MANIFOLD-5 (we had run out of cartoons and needed something to fill up the space - an obituary sounded novel).   He has an uncanny knack of following Ian Stewart around and has appe-ared to date in Open University course text margins, a cookery book called *Simple Scoff*, the *Mathematical Intelligencer*, and a paper-back called *Nut-Crackers*.

MANIFOLD even had a constitution.   The most important rule was that there had to be an AGM once a year, and *it had to be held in a pub*.

The cover changed colour each issue.   We began with orange, be-cause that's what the Maths Inst. used for preprints, and there wasn't anything else.   We moved to green, which we *bought*, to avoid being orange forever, as Springer Verlag is indelibly yellow.   Issues 3-8 were all on quarto paper, but had graduated to multilith instead of Roneo: yellow, blue, pink, yellow, pale green, dark blue.   At issue 9 international paper sizes had come, and we became A4, changing printers to the University of Nottingham who were cheaper.   The colours contin-ued: orange, blue, grey, pink, dark green, pink, blue (reprinted yellow), purple, pale green, orange.   But by now the energy was running out.

MANIFOLD

Autumn 1975

17

THE ECONOMICS
OF SLEEPING

MANIFOLD *seventeen:*
Ersatz Pooh in sickly green.

The schedule was three issues a year: by 1979 we had only reached issue 19 (back at Warwick, in blue), and the flow of letters read-ing "Dear Sir, I have not received MANIFOLD for seven years; I pre-sume my subscription has run out - I enclose $20" was growing stea-dily.   And so, in 1980, in eyeboggling day-glo lime green, Fast-print Ltd. of Streatham Hill, London, saw us to our rest.

4

The original editorial board consisted of John Jaworski, Ramesh Kapadia, Donal Monaghan, Ian Stewart and (no relation) Mark B. Stewart. After issue 11, Robin Fellgett joined and promptly invented the single ploy that MANIFOLD holds most dear: the Overseas Institutional Subscription, which probably saved it from bankruptcy.

MANIFOLD enriched all our lives. We recall the joys of stapling 600 copies per issue, using a curious device with a big lever and a knob that kept falling off. When we got fed up we let Nottingham do it for an extra £5. This led to the termly dash up the M1 to load boxes into the boot before they got lost.

There was the one and only MANIFOLD Annual Dinner at the Westgate arms, Warwick: a meal for nine with all the trimmings for £40, marred only by a rapid trip home to get ties and jackets. There was the Turkish gentleman with his new theory ("...I felt dangerous part of the theory of gravity... I am glad... to complete the theory.") There was William L. Fischer who sent us proofs of the inconsistency of non-Euclidean geometry (with Euclidean geometry, which seemed fair enough to us!). There was our first unsolicited letter: "Dear Sirs, would it be possible to number the pages?" (We did, thereafter, except for issue 20, when Ramesh forgot.) There was the annual sale of MANIFOLDs at the British Mathematical Colloquium: we left piles on a table, and let people put money in a box. We once sold £43 worth in an afternoon (at 15p. each). There were the early attempts to become rich and famous. Our stab at the first was to court the motor industry for support (offering to call our rag ROOTS, after the Rootes firm in Coventry). In view of later developments at Warwick, this was probably a narrow escape. Our stab at the second was to encourage banning in the USA, by sending free copies to Cuba. Neither country seemed especially concerned.

And that brings us to our final unwillingness to let go. The idea of a 'Best of MANIFOLD' collection has been with us since the early 1970's; but it has taken until the 80's, with the courage and foresight of Biga Weghofer, for this volume to appear. This distillation from the seven years' worth of MANIFOLD published between 1968 and 1980 is in no sense a *best* of MANIFOLD, however. We spent a wintry Sunday re-reading from our back numbers to compile this celebration. It was apparent from the start that some of MANIFOLD could never be included, simply because mathematics had not stayed still. Among our more serious and successful features was 'Mathematics of the Seventies', which surveyed over a period of two years exactly where the frontiers of mathematics had got to. They have now moved on, and the period charm of those items is not enough to counter their irrelevance. We have also decided not to include anything from our two full issues devoted to Catastrophe Theory.   That too has not stood still, nor would a selection adequately represent the wide range of ideas that filled those issues.

And so, in the following pages, you will find a concentrated dose of the 'spirit of MANIFOLD'.   We have corrected the misprints, invented new ones, done a certain amount of internal editing where necessary, provided the briefest of commentaries and then left you to yourselves.   The shades of Jozef Plojhar and Eve laChyl, who have gone to the great printing-press in the sky, watch over you!

J^2 + INS
Potterspury and Barby 1981.

SIGNS

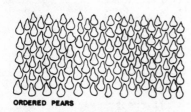
ORDERED PEARS

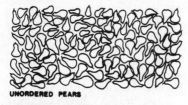
UNORDERED PEARS

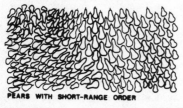
PEARS WITH SHORT-RANGE ORDER

M-20

For any Mediaevalists we may have among our readers, a quote from Prof. D.J.Hinton, Professor of Architecture at Aston University: "If people can really be divided into two categories they are (a) Those who think people can be divided into two categories, (b) Those who do not."   M-4.

The penchant of certain literary and artistic figures for elaborate cryptograms and cyphers is well known. Schumann used a musical cypher, and it is widely believed that Shakespeare's dedications contain the secret of "Mr. W.H."   Another confessed cryptographer was Elgar; his code has always defied analysis until recently, when Eric Sams published some solutions.   The one most likely to attract popular attention is the surprising fact that the Enigma Variations are variations on no less a theme than "Auld Lang Syne".   This seems to be well substantiated, as the same analysis reveals that ENIGMA is a coded notation for the opening bars of the piece; and that EDWARD ELGAR gives one of the characteristic melodies: a fact supported by the observation that Elgar signed some of his works with the musical notation for that melody.   M-6

In the same vein: whose musical cypher is this?   See below for the answer.

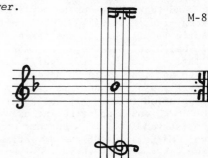

M-8

BACH.   The notes, in rotation, are B,A,C,H (German notation).

*Bourbaki. Who are these Frenchman, and why aren't
he all French? Ungrammatical, but logical: just
like Bourbaki (t)hi(e)mself(ves). An analysis text
has the index entry: BOURBAKI - see LANGUAGE, ABUSE
OF. Quite so. Of mathematics too, some say, but
that's just abuse...*

BOURBAKI

ELIZABETH CAMPBELL

Mention Bourbaki to some mathematicians and you will encourage a
lengthy discussion; to others and you will be answered with "who?"
or "what?". So for those in the second category an introduction
to Bourbaki is the place to start. Bourbaki is a name, in full
"Nicolas Bourbaki"; however this Bourbaki is not a person, it is a
pseudonym for a group of people. This group of mathematical writers started bringing out publications in the mid 1930's. The
group of course does not have the same members as at its foundation,
several having resigned and new recruits having joined.

Nobody is quite certain why the group adopted the name "Bourbaki".
The most plausible theory is that it was inspired by a soldier of
the mid 19th Century who was of some importance in the Franco-Prus-
sian War - General Charles Denis Sauter Bourbaki. This explains
the surname but not the "Nicolas", and the theory would not hold
much water except that there is a statue of him in Nancy and that
several members of the group were, at one time or another, associated with the University of Nancy. The General seems a likely character for a piece of fiction; for in 1862 at the age of 46 he turned down a chance to become King of Greece; and in 1872, after

*an invitation to the
marriage of Mlle.
Betti Bourbaki and
M. Hector Pétard*

Monsieur Nicolas Bourbaki, Membre Canonique
de l'Académie Royale de Poldévie, Grand Maître
de l'Ordre des Comparts, Conservateur des
Uniformes, Lord Protecteur des Filtres, et
Madame, née Bitwitvoque, ont l'honneur de vous
faire part du mariage de leur fille Betti avec
Monsieur Hector Pétard, Administrateur-Délégué
de la Société des Structures Induites, Membre
Diplômé de l'Institute of Class-field Archaeo-
logists, Secrétaire de l'Œuvre du Sou du Lion.

Monsieur Ernst Stanislas Pondichéry, Com-
plexe de Recouvrement de Première Classe en
retraite, Président du Home de Rééducation des
Faiblement Convergents, Chevalier des Quatre U,
Grand Opérateur du Groupe Hyperbolique,
Knight of the Total Order of the Golden Mean,
L.U.B., C.C., H.L.C., et Madame, née Compactensol,
ont l'honneur de vous faire part du mariage de
leur pupille Hector Pétard avec Mademoiselle
Betti Bourbaki, ancienne élève des Biscordonnées
de Bosse.

L'isomorphisme trivial leur sera donné par le P. Adique, de l'Ordre des
Diophantiens, en la Cohomologie principale de la Variété Universelle, le
3 Cartembre, an VI, à l'heure habituelle.

L'orgue sera tenu par Monsieur Modulo, Assistant Simplexe de la Grassman-
nienne (lemmes chantés par la Scholia Cartanorum). Le produit de la quête
sera versé intégralement à la maison de retraite des Pauvres Abstraits. La
convergence sera assurée.

Après la congruence, Monsieur et Madame Bourbaki
recevront dans leurs derniers fondements. Sauterie
avec le concours de la fanfare du 7e Corps Quotient.

Tenue canonique
(idéaux à gauche à la boutonnière).   C. Q. F. D.

leaving France with the remains of his army, he tried to shoot himself while imprisoned in Switzerland. He must have been the worst marksman in the army, for he is reported to have died in 1899 at the age of 83.

Another story of the origin of the name, probably a rumour started by the group themselves, is that about 35 years ago, at the Ecole Normale Superieure, annual lectures were given to first-year students by a visiting speaker, who was in fact an amateur actor in disguise, and whose lecture was just a lot of mathematical doubletalk. This visiting speaker went by the name of Nicolas Bourbaki...

This story is typical of the group. They have no oath of secrecy, but do not like their secrets to be known. The pseudonym is used as a corporate name rather than as a disguise, and the names of most members are known to many mathematicians. The membership seems to vary between about 10 and 20. Originally it was completely French; but nowadays distinguished mathematicians of many nations are included. Michael Atiyah of Oxford is a notable English member. The first non-French member was Samuel Eilenberg, a Polish mathematician at Columbia University who knew more about algebraic topology than any Frenchman. The Frenchmen-only rule was waived, and Eilenberg joined the group.

Bourbaki's first publications appeared in *Comptes Rendus*, a journal of the French Academy of Sciences. In 1939 he began to publish his *Elements of Mathematics*, and it is this treatise which attracts the most attention. It is a comprehensive study of mathematics starting with the most basic principles. An explanation of the treatise was given in the *American Mathematical Monthly* in 1950 under the title 'The Architecture of Mathematics'. A footnote to the article reads:

> Professor N.Bourbaki, formerly of the Royal Poldavian Academy now residing in Nancy, France, is the author of a comprehensive treatise of modern mathematics, in course of publication under the title *Eléments de Mathématique*. (Hermann et Cie., Paris 1939).

Twenty-six volumes have appeared to date, volume XXVI being *Groupes et Algèbres de Lie* published in 1960. The volumes are published by Hermann (in paperback), and do not appear in the same order as the divisions of the treatise specified by Bourbaki - but they give the book and section numbers as in the treatise (which is confusing). Translations of some of the volumes published by Addison Wesley can be obtained in hardback.

The way the treatise is written is of as much interest as the object itself. Because the treatise starts from basic principles, each book assumes only what has been proven in previous books. Any example which refers to facts which may be known to the reader, but have not yet been proved in the text, is put between asterisks. All volumes contain many examples and exercises. Other peculiarities are the use of a sign in the margin ("dangerous turning") to indicate a passage warning the reader against a standard error; the use of small type for passages that can be omitted on first reading; abbreviations with a particular meaning in a limited section of the

UNIVERSITÉ DE PARIS
—
ÉCOLE NORMALE SUPÉRIEURE

45, Rue d'Ulm, Paris-5⁰
Tél.   Paris, le 24 octobre 1967

Messieurs les Professeurs E. C. ZE
et H. REITER
Mathematics Institute
University of Warwick
Coventry

Mes chers Collègues,

Je vous remercie vivement de votre invitation. Elle était très aimable
et très flatteuse.

J'aurais aimé participer à votre symposium qui promet d'être fort
intéressant. Mais ma timidité et ma modestie bien connues m'interdisent de
parler en public. D'autre part, avec l'âge, je répugne de plus en plus à me
séparer de mes collaborateurs, et ne puis en déléguer aucun.

J'envoie mon cordial souvenir à tous les congressistes, et souhaite
que le symposium fasse naître beaucoup de théorèmes.

N. BOURBAKI

Christopher Zeeman and Hans Reiter invited Bourbaki to attend the Warwick University Harmonic Analysis Symposium in 1968. This was his reply.

treatise; and another symbol to mark difficult exercises.

Bourbaki brought out other publications. One, *Foundations of Mathematics for the Working Mathematician* appeared in the Journal of Symbolic Logic for 1949, giving the author's home institution as the University of Nancago. This name is built from Nancy and Chicago - the latter because one of the founders, Andre Weil, had moved there. "Nancago" also appears in a series of advanced mathematical books under the collective heading *Publications de l'Institut Mathématique de l'Université de Nancago.*

Bourbaki, for all his impressive publications, could not gain membership of the American Mathematical Society. Its officials rejected his application and the secretary of the society suggested he apply for Institutional Membership. The same secretary appears in another tale about Bourbaki. In the late 1940's a paragraph about Bourbaki appeared in the Book of the Year of the *Encyclopaedia Britannica* , describing him as a group, and written by Ralph P. Boas, then editor of *Mathematical Reviews*. Soon after, the editors of the *Britannica* received an injured letter, protesting against Boas's allegation that Bourbaki did not exist, and signed by the gentleman themself. The editors' confusion and embarrassment were heightened by a truthful but deceptively worded letter implying that Bourbaki did in fact exist. Then the secretary of the American Mathematical Society decided to clear the matter up by writing to the editors. But Bourbaki remained one up, and started a rumour that Boas did not exist: that in fact Boas was the collective pseudonym of a group of young American mathematicians who acted jointly as editors of *Mathematical Reviews*.

This illustrates the youthful vitality of the group, despite the fact that some members remain so until aged about 50. The real attraction of Bourbaki, however, is not in the humour of his authors, but in the uniqueness of his treatise. It gives the first systematic account of some subjects, and of mathematics in full, not available in any other publication.

9

This symbol, inscribed upon coins found at Knossos, is thought by some hopefuls to be a map of the infamous Labyrinth of the Theseus-Ariadne story. Credulity is strained: one feels it should have been more challenging - enough to inspire:

a new Introduction to Maze Theory

STEVEN EVERETT

Theseus and the Minotaur are to be seen sitting on rocks opposite each other in the heart of the Labyrinth. Having found a companion of no mean intelligence, the Minotaur (bored stiff) is not going to waste him as mere nourishment, while Theseus has no option - his fabled ball of string was about as long as all other balls of string, and gave out well before the second junction of the Labyrinth. The conversation has exhausted all the usual mathematics of mazes, left hand on wall methods, Eulerian unicursality, and so forth. (For anyone who doesn't know about all these, and those for whom this whimsical piece is a *real* introduction to maze theory, Rouse Ball's *Mathematical Recreations and Essays* will give you the best grounding.)

Mino: You know, Theseus, in time they will have machines capable of threading labyrinths - I hope they name them after one of us!

Thes: I don't know, though - most of the fun of this sort of myth would be lost if we could get out of here: sort of keeps the mystery up, you know? If we ever got out of here, all we could do is sell our story to the *Weekend Heliograph* magazine, or VARIFOLD possibly, and then in a couple of weeks all the excitement about mazes would be over.

Mino: Yes, I see what you mean.

Thes: No books about mazes, mathematical magazines deprived of half their contributions...

Mino: I've got a small work here that I wrote a century or two ago, called 'Three Minotaurs in a Boat' - it's got a very funny chapter about mazes...

Thes: Yes, I must read it some time.

Mino: Aren't there any *other* theorems about mazes that you mathematicians have brought up? That 'left hoof on wall' method always confuses me, as I've got two left hooves, and I'm not too sure about which hoof I started with - besides, it might not work here, you know.

Thes: There is one little-known theorem about mazes, you know - if I can only remember how it goes.

Mino: Do try to think, Theseus, please - it's frightfully dull in here!

Thes: As I remember, it goes something like this - there exists a sequence of left and right turns which will get you out of *any* maze...

Mino: You mean a sequence like LLRLRRRLLR...?

Thes: Yes, that's right - I don't know the sequence, mind you!

Mino: Have they actually worked the sequence out?

Thes: No, I don't think so - the theorem only tells you that there *is* a sequence, it doesn't help you find it; it's what they will be calling an *existence proof* in Twentieth century mathematics.

Mino: Can't you remember some of the proof?   We've got a lot of time, perhaps we could work something out?

Thes: Well, I know that to start with you must make a list of all the possible mazes - start with the very simplest one and call it number one...

Mino: What is the simplest maze?

Thes: A T-shaped one, I suppose - one with a junction, one entrance, and two blind alleys.

Mino: Isn't a straight line a simpler maze? An I-shaped maze?

Thes: I suppose so, but I don't think it's important.

Mino: What about a blind alley, though?   You can't make a right or left turn at the end of a blind alley, can you?

Thes: No, but I think you have to agree to turn back from blind alleys, I'm sure you do!

Mino: What about junctions with more than three passages - ones where you can go 'half-left' or straight on?

Thes: That's all been sorted out, I'm sure - I can't remember the exact details, but don't bother with that now, I'm beginning to get myself interested.

Mino: Yes, yes!   Go on!

Thes: Well, when you've listed them all like this, you look at the first one and choose a place in it and work out the sequence

11

of right and left turns to get you out from that point.
Then you choose another place and see where the previous sequence would have got you to if you hadn't been in the first
place...

(There is a long pause while *Mino* digests this - Theseus is
about to speak again when, surprisingly, Mino interrupts.)

Mino: And then I suppose you tag on to the first sequence a new sequence to get you out from there?
Thes: Yes, that's right!
Mino: And then go on to do it for all the other places in the
first maze?
Thes: Yes, and then you go on to the second maze, and then the third, and so on.

(*Mino* thinks a little before asking slowly:)

Mino: There are an infinite number of positions in these mazes and
an infinite number of possible mazes, aren't there?
Thes: Yes, but we can still make a list of them, which is what matters: there are an infinite number of them, but a *countably*
infinite number, which makes it all right mathematically -
the left-right sequence is going to be infinite, you know.
Mino: Yes, Theseus, I see that. But if I *can* count this infinite
set, I can count them in any order, can't I?

(*Thes*, sad to relate, is a very bad mathematician, and hasn't
noticed that this proof works only if mazes are counted in
order of complexity - or else has failed to distinguish between countable *cardinals* and *ordinals*. Poor Theseus.)

Thes: Yes.
Mino: Well then, can't I count them so that all the mazes which
take *left* turns to get out of come first? So that my sequence begins with an infinite number of left turns: LLLL...

(*Thes* doesn't reply, which is a pity: *Mino* runs off into the
Labyrinth and his voice is heard getting fainter and fainter)

Mino: Left.........left.....................left.......

(*Thes* is also a slow mathematician. It is fully thirty sec-
onds before he rushes off after the Minotaur, calling to himself:)

Thes: Right.........right.....................right......

After an hour or two of such frivolity, Theseus stops turning
right and looks carefully around to make sure he is not observed,
before gingerly placing his left hand on the wall and setting off.
After a while he passes the Minotaur, limping along with both left
hooves on the opposite wall, travelling in the opposite direction.
Both look highly embarrassed, and pretend to be deep in thought,
and oblivious of any other people, as they pass.

One of MANIFOLD's *first 'breakthrough' articles*   M-3
reported an important discovery in topology: the
long-sought proof of

the Annulus Conjecture

BARBARA SANDS

Topology, the saying goes, is a sort of geometry. It differs from
ordinary geometry in that it considers many things to be 'the same'
which would ordinarily be considered different. More specifically,
two topological spaces are said to be *homeomorphic* (or *topologically equivalent*) if one can be transformed in a nice, continuous sort
of way into the other. For example, the circular disc and the square in the plane are homeomorphic: we just round off the corners.
This is best done by shrinking radial lines from the centre of the
square until they are all the same length. So far as topology is
concerned, the disc and square are equivalent; each is called a
2-dimensional ball, or *2-ball*. The n-*ball* is defined similarly to
be anything homeomorphic to the n-dimensional disc

$$x_1^2 + x_2^2 + \ldots + x_n^2 \leq 1$$

in n-dimensional space; and by similar reasoning the n-cube
$|x_1| \leq 1, \ldots, |x_n| \leq 1$, is also an n-ball.

   If an n-ball B contains a subset C homeomorphic to an n-ball B',
we say that B' is *embedded* in B. The sort of thing we think of
is:

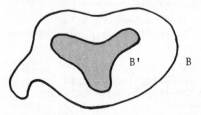

where the shaded part is just a squashed-up disc, and we get a
2-ball embedded in a larger 2-ball.
   A very entertaining example of an embedding of 3-balls is the
Alexander horned sphere. It is constructed in stages from a
standard 3-ball. First two horns are pushed out. Each horn is
split in two and the ends are intertwined; then the new horns are
split in two and intertwined, and so on. Repeating the construction indefinitely we obtain as a limit a rather complicated object.
However, topologically it is just a 3-ball: the method of extruding
horns defines, in the limit, a homeomorphism between the initial
3-ball and the final result. The horned sphere is embedded in

3-dimensional space by construction: we just restrict things a bit
and embed it in a 3-ball:

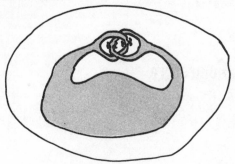

Now consider the space between two balls, one embedded in the
other. If we put one standard ball inside another, with their
centres coinciding, the space in between is called an *annulus*, and
is just a hollow ball with a thick 'skin'.

We can parametrize points in the annulus by pairs (x,d) where x
is a point on the boundary and d is a radial movement inwards. Twi-
sting things about a bit doesn't alter much: we can parametrize the
same way using the twisted 'radial lines':

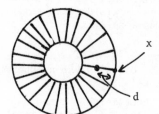

This picture suggests the *Annulus Conjecture*:

> For any 'nice' embedding of an n-ball in another
> n-ball, the space between them is homeomorphic
> to an annulus.

Before explaining what 'nice' means (the point is that the conjecture is *false* for arbitrary embeddings) let's look at the horned
sphere again. Put a loop L round one horn. Then it's clear (and

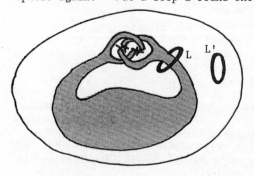
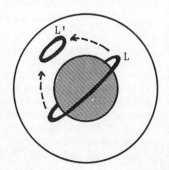

provable) that we can't pull L off the horn without breaking it,
to a position L', because it gets 'tangled up' in the horns.   On
the other hand, for the usual embedding of two standard 3-balls,
any loop L *can* be pulled off, as in the right-hand diagram above.

Since the possibility of pulling loops off is preserved by home-
omorphisms, it follows that the space between the horned sphere and
its surrounding 3-ball is *not* an annulus, and the conjecture fails
in this case.

So whatever 'nice' means, the embedding of the horned sphere is
decidedly *nasty*.   The trouble, one suspects, arises because the
embedding is so tangled up.   If we consider a point right at the
'tip' of the horns, then the embedding looks horribly messy:

no matter how closely we look.   This suggests that 'nice' is what
topologists call *locally flat*: near a boundary point of the inner
n-ball, it should look like this:

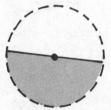

For many years people tried to prove the Annulus Conjecture for
locally flat embeddings, but without success.   The situation is
now radically different.   The young American mathematician Robion
Kirby (Los Angeles) provided the main ideas that led to the breakthrough, by reducing the problem to a much more special one.   It
transpired that Terry Wall (Liverpool) had already solved the special problem, but not published the proof!   Between them, Kirby and
Wall had cracked the conjecture.   The majority of the credit, however, must go to Kirby.

Thus we may state the *Annulus Theorem*:

> The Annulus Conjecture is true for locally flat
> embeddings of n-balls, provided n $\neq$ 4.

(There is something very funny about 4-dimensional space, and very
nearly everything in topology is unsolved in dimension 4.   The re-
striction n $\neq$ 4 came as no surprise.)

Why is the annulus theorem important?   One of the long-term
aims of topology is to classify all possible *manifolds*.   A space M
is a manifold provided that near each point it looks like Euclidean
space.   If it looks like Euclidean n-space it is an n-manifold.
For example the 2-*sphere* S^2 and the 2-*torus* T^2 are 2-manifolds, and
we can draw a typical point with a little bit of 2-space around it:

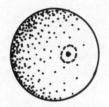

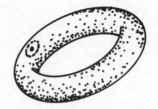

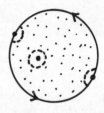

2-*sphere* S^2 2-*torus* T^2 *projective plane* P^2

Less obvious, but important, is the *projective plane* P^2 obtained from a disc by 'glueing together' diametrically opposite points. It can't be drawn very convincingly because it won't embed in 3-space without crossing itself.

It turns out that it is possible to build up all 2-manifolds (surfaces) from these three types by sticking them together using the *connected sum* #. If M, N are manifolds we get the connected sum M#N by removing small discs from M and N and fitting a 'tube' from one hole to the other. So

S^2#T^2 looks like: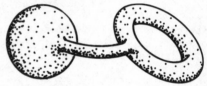

The highly important *Classification Theorem for 2-manifolds* says that any 2-manifold (assumed compact) is homeomorphic either to

$$S^2\#\underbrace{T^2\#T^2\#\ldots\#T^2}_{n} \qquad (n \geq 0)$$

or to $$S^2\#\underbrace{P^2\#P^2\#\ldots\#P^2}_{n} \qquad (n \geq 1)$$

and all these, as n varies, are different. For a sketch proof, stated in slightly different terms, see [1].

If we try to do the same thing for n-manifolds, we find that we can define the connected sum the same way. Or can we? I've cheated a bit by calling it *the* connected sum: how do I know that it does not depend on whereabouts I cut out the holes? For 2-manifolds everything's OK, by the classification theorem; but what about n-manifolds?

That's where the Annulus Theorem comes in - as a first step towards classifying n-manifolds. What we need to prove boils down to this: if D and D' are two discs embedded in a manifold M, then we can push D around until it coincides with D'. The way to start is to shrink D until it is very small, and then slide it across M until it gets inside D'.

But now the Annulus Theorem lets us draw 'radial' lines in the space between D and D', and we can expand D along the radii to make the two discs coincide.

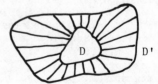

16

So the Annulus Theorem proves that M#N is well-defined in n-space.

Finally, the proof. It is extremely long, and uses most of the topological machinery available: a list of all discoverers of results used in the proof would include all major topologists of the present century!

The way we shrunk D to get it inside D' above suggests that to prove the theorem, it should be sufficient to consider only a tiny patch of the embedding. This means that something deeper than properties of balls is involved; and it turns out that what matters is the *PL-structure* induced by the embedding - a kind of triangulated grid structure on the space, like a polyhedron.

Kirby's brilliant idea was to 'embed' a standard manifold T^n, the n-*torus*, in the n-ball B^n - but allowing the torus to 'overlap' itself (technically it is an *immersion*, and a small disc is removed from T^n first). This immersion of T^n induces the same PL-structure as the embedding of the ball, and we now have the problem of uniqueness of PL-structures on T^n. Now T^n happens to be a very nice manifold for a construction known as *surgery* (cutting and glueing in n-space) and the surgery problem is what Wall had solved. Surgery is a useful tool for problems of uniqueness of PL-structure, and although Wall had not worked out all of the details for T^n, he had enough information to complete the proof of what, henceforth, must be named the Annulus *Theorem*.

BIBLIOGRAPHY

1. Ian Stewart   *Concepts of Modern Mathematics*, Penguin 1975.

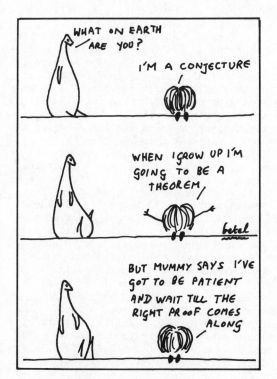

M-10

betel

*You can lie about statistics, you can lie with
statistics.   But did you realise that it is
possible to use valid statistics to get a correct answer... and still lie?*

Genuine Fakes

BARRY PILTON

Throughout history there has been one major problem facing scientist and mathematician alike.   I am referring not to Squaring the
Circle, but to the more practical problem of Cooking the Books.

Faking statistical results is a hazardous enterprise; but there
are times when observations and theory disagree, and the only reco-
urse is fiddling the results (or similar viol practices).   The
great thing about statistical analyses is that they can be carried
out years after the figures are produced (indeed this is customary
in government circles), and advances in statistical technique can
sometimes reveal faking where no faking was previously observed.

The 'fit' of experimental theory with actual figures, as well as
being too bad, *can also be too good*.   If you throw 1000 coins and
claim to have got exactly 500 heads and 500 tails, don't claim it
too often: the probability is less than one in a hundred.

A remarkable instance is that of Mendel, whose experiments with
plants (the good man was a monk and so confined his genetic experiments to botany) form the foundation of modern theories of heredi-
ty.   Taken as a whole, Mendel's results are too good to be true.
No one is suggesting that he knowingly faked them; but it does seem
that when he was deciding whether a particular plant was 'dwarf' or
'tall', his judgement was not entirely impartial.   Then there was
a man called Moewus who counted algae, getting results agreeing so
well with theory that, had the experiment been repeated by the
whole human race every day for ten billion years, as good a fit
might happen once.   It may have been just luck, but it's not the
sort of luck to put into a Ph.D. thesis.

So scientists fake their figures: we mathematicians have always
suspected as much.   But what of mathematical fakery?   Not much
chance there... but indeed there is: the very famous experiment by
Lazzerini to evaluate π by dropping needles on to ruled lines.
This has been reported by various eminent authorities with a smidg-
in of over-respect.   That Lazzerini was a successful hoaxer has
been shown by O'Beirne [4] and Gridgeman [2], independently.

The theory is due to Count Buffon.   Drop a needle of length d
on to a grid of parallel lines distance a apart, where a > d.
What is the probability that the needle lands across a line?   The
answer is $2d/\pi a$, so experimental evaluation of the probability as a
proportion of successful trials can yield an estimate for π.   In
1901 Lazzerini made 3408 tosses, and obtained $\pi \sim 3.1415929$, in error by about 0.0000003.   This is remarkably close, but until

recently only one authority appears to have smelt a rat.   In [1]
p.176 Coolidge says: "...it seems quite likely that Lazzerini
'watched his step' and stopped his experiment at the moment he got
a good result."

In the wake of these doubts come many more.   If the result is
accurate to one part in a million, presumably so were the measure-
ments of d and a.   If not, the result is presented to ridiculous
accuracy.

...and in only 3408 trials...?

Suppose there are s successes in n trials.   The estimate is $\pi \sim$
2dn/as.   The ratio d/a is most likely to be chosen as a fairly simple fraction, so this expression gives a good approximation to π in
rational numbers.   A good approximation is 355/113: the next better
is 52163/16604.   And 355/113 = 3.1415929... which is Lazzerini's
result.   Further, the unusual prime 71 divides not only 355, but
also 3408.   Hence if d/a is a fraction with numerator 5, the most
likely denominators are 6 or 8, and Lazzerini's results would be
obtained if the number of successes were 1808 or 1356.

So O'Beirne reasoned: then he looked up Lazzerini's original paper [3].   The first discovery was that his name is actually
Lazzarini, so everyone else had been copying each other.   The next
was that d/a *was* 5/6; and the number of successes indeed 1808.   It
was then obvious what had happened.   Arrange for d/a to be *near*
5/6, and assume this figure exact.   Count your successes for blocks
of 213 trials, *continue until the number of successes is the same
multiple of* 113, then stop.   You ought to get equality somewhere
before 5000 trials.

But Lazzarini's fakery doesn't stop there.   He also lists results after various numbers of trials; again the agreement with theory is far too good.

For those who still think this method is worth trying: to have
confidence in the first n figures for π you need to make roughly
10^{2n+2} trials.

BIBLIOGRAPHY

1. J.L.Coolidge   *The Mathematics of Great Amateurs*, Dover 1963.
2. N.T.Gridgeman   *Geometric Probability and the Number* π, Scripta
   Mathematica 25 (1960).
3. M.Lazzarini   *Un' applicazione del calcolo della probabilita al
   ricerca sperimentale di un valore approsimato di* π, Periodico di
   Matematica (1902).
4. T.H.O'Beirne   *Puzzles and Paradoxes*, Oxford University Press,
   1965.

* * * * * * * * *

Legendre, examining entrants to the École Polytechnique.   The
student: Arago.   A question involving double integrals.

L: That's not how you were taught to do it.
A: No, I found it in one of your memoirs.
L: Why did you use it?   To butter me up?   M-17
A: Nothing was further from my mind.   I adopted it
   because it seemed preferable.
L: If you cannot explain to me *why* you think it preferable, I
   shall be forced to fail you.

SIGNS

Stan Kelly, folksinger and computer consultant, is perhaps best known for his Liverpool Lullaby, recorded by The Spinners. In 1968 he was a graduate student at Warwick, and combined his talents with:

Green Grow the Fibres~O!

STAN KELLY

The drunken mathematician is not a pretty sight, but then, who is? As the glasses raise on high, and the evening grows merrier, the following noise may be heard.

Smooth flow on the Manifold-O!   (Or: Green grow the Fibres-O!)

I'll sing you ONE-O!   Green grow the fibres-O!
What is your ONE-O?
ONE is my identity, non-trivial to me.

I'll sing you TWO-O!   Green grow the fibres-O!
What are your TWO-O?
TWO,TWO, the circular points, way out at infinity;
ONE is my identity, non-trivial to me.
    and so on, using:

THREE, THREE, the old triangle...
FOUR for the colours on the map...
FIVE for the genus of my vest...
SIX for the perfect number...
SEVEN for the sides of a ten-bob note...
EIGHT for the Ceilidh (Cayley) Numbers...
NINE for the nine-points circle...
TEN for the base of common logs...
ELEVEN for the fifth prime number...
TWELVE for the eleven-plus...
Lemma Three

MARGINAL NOTE

Chorus: Lemma 3 very pretty, and the converse pretty too;
        But only God and Fermat know which of them is true.

When I studied Number Theory, I was happy in my prime;
And Fermat's wild conjectures, I knocked them two at a time. [Ch.]

Black and white together, we shall not be moved;
But the 4-colour Theorem, it hasn't yet been proved. [Ch.]

Now Lemma 3 has puzzled mathematicians by the score,
But Max Newman has engulfed it and it won't be seen no more. [Ch.]

The axioms of Choice are very clear to me:
If you want to choose a Lemma, man, well don't choose Lemma 3. [Ch.]

* * * * * * * * *

Prof. M.H.A.Newman was Stan Kelly's research supervisor.

Exponential population growth is generally credited   M-11
to Thomas Malthus in 1798.   Actually, he was anti-
cipated by some six centuries.   Moreover, the ear-
lier work has potential applications to computing...

the Good-natured Rabbit-breeder

JOZEF PLOJHAR

The breeding of rabbits has brought us a long way.   In 1202, Leonardo of Pisa (who had nothing at all to do with anything called
the Mona Lisa) was formulating a theory of rabbit-breeding.   He
observed that if one begins with a pair of new-born rabbits, if one
assumes that rabbits become productive in the second month, and if
one assumes that thereafter each productive pair gives birth to a
further pair each month, then the total size (in pairs) of the rabbit colony was given by what is now called the *Fibonacci Series*:

$$1 \quad 1 \quad 2 \quad 3 \quad 5 \quad 8 \quad 13 \quad 21 \quad 34 \quad 55 \quad 89 \quad 144 \ldots$$

At least, as long as no rabbits die.

Leonardo, in picturesque Pisan fashion, was called 'son of good
nature', which is *fibonacci* in Italian, and is not to be confused
with that well-known mathematical Mafia, The Bernoullis.

In the Fibonacci series, each term is the sum of the two prece-
ding.   This difference equation can be solved by traditional methods, introducing the constant

$$\phi = \tfrac{1}{2}(1+\sqrt{5}) = 1.6180339887498948482045868343656381\ldots .$$

The nth Fibonacci number $F(n)$ is given by

$$F(n) = (\phi^n - (-\phi)^n)/\sqrt{5} = [(\tfrac{1}{2}(1+\sqrt{5}))^n - (\tfrac{1}{2}(1-\sqrt{5}))^n]/\sqrt{5}.$$

The number ϕ appears in many unusual places.   It is the Golden
Ratio, beloved by mathematical aesthetes; it has the interesting
property of being 1 greater than its reciprocal; and it is the limit of the ratios of successive terms of the Fibonacci series.   The
series itself turns up in the botanical study of phyllotaxis - the
arrangement of leaves and branches around a central stem.

Many of the more spectacular properties of the Fibonacci series
depend on the following, easily proved property: every positive integer N has a *unique* representation

$$N = F(k_1) + F(k_2) + \ldots + F(k_r)$$

where $k_i \geq k_{i+1}+2$.

The proof hinges on the fact that the only possible choice for
$F(k_1)$ is the largest Fibonacci number not greater than N.   We can
then apply the same argument to $N-F(k_1)$, and so on.

One of the more surprising and less well-documented properties
of ϕ is the following: any positive integer can be expressed as the
sum of a finite number of distinct integral powers of ϕ.   To

WINTER 1971-72 **manifold**

the good-natured
rabbit-breeder

provide supporting evidence for this assertion, let us give some examples:

$$1 = \phi^0$$
$$2 = \phi + \phi^{-2}$$
$$3 = \phi^2 + \phi^{-2}$$
$$4 = \phi^2 + \phi^0 + \phi^{-2}.$$

A proof is reasonably easy, given a little direction in the matter of which properties of Fibonacci series to look at. We might begin, for example, with the representation above, coupled with the additional result that

$$F(n) = \phi^{n-1} - \phi^{-1}F(n-1).$$

However, a good mathematical principle is never to bring up heavier artillery than we need, and it turns out that we can get well along the road to the same result by means of the much simpler observation that

$$N = 1 + 1 + 1 + \ldots + 1 = \phi^0 + \phi^0 + \phi^0 + \ldots + \phi^0.$$

So far, this is what we are trying to prove, with the failing that the powers do not appear with the coefficients 1 and 0 only. A general proof that we can reduce a case such as this to the desired result might be rather confusing, so let us look at a specific case, say at $4 = 4.\phi^0$. We deal with this by noting that

$$\phi^{n+1} = \phi^n + \phi^{n-1} \qquad\qquad (*)$$

which follows since $\phi = 1 + (1/\phi)$. Rearranging this we get, after a little manipulation,

$$2\phi^n = \phi^{n+1} + \phi^{n-2}.$$

Hence we get

$$4 = 2(\phi + \phi^{-2})$$
$$= 2\phi + 2\phi^{-2}$$
$$= \phi^2 + \phi^{-1} + \phi^{-1} + \phi^{-4}$$
$$= \phi^2 + \phi^0 + \phi^{-3} + \phi^{-4}.$$

This is satisfactory, and a little checking with pencil and paper should convince those who distrust the manipulations. However, this is not the representation of 4 given above. The expansion exists, but is not unique! As an exercise: derive one form from the other using (*) above.

Thus we have pointed to a proof that any positive integer can be represented as a *number to base* ϕ. Looking at things the other way round, we can imagine that we are dealing with numbers such as 101.01 - sequences of position-dependent 0s and 1s. This corresponds in a fairly obvious way to the representation of 4 that we found above:

$$101.01 = 1.\phi^2 + 0.\phi + 1.\phi^0 + 0.\phi^{-1} + 1.\phi^{-2}.$$

We shall find this a convenient way to write such numbers, partly to avoid layout difficulties on the printed page, but more importantly because such a representation contains all that we need to know about the numbers we are discussing. In contexts where we might be confused we could emphasize this by writing 101.01_ϕ.

Sometimes mathematical tails wag mathematical dogs. This is just such a case. Once we begin to think of *Fibinaccary* numbers (there are many possible names, but few have the resonance of this!) we make some natural comparisons with the regular binary (base 2) representation - in which 101.01_2 corresponds to 5.25_{10} , and 4_{10} corresponds to 100_2. In both cases, binary and Fibinaccary, we have a means of representing an integer by a finite string of 0s and 1s. The relevant practical application is, of course, computing. In the binary digital computer a collection of bi-stable devices (capable of recording or registering either of two states: on/off; +/-; 0/1) is associated by the design of the computer to store a full integer. Interest in binary notations arises because bi-stable devices are easier to come by than devices capable of distinguishing between a greater range of values.

There are some problems to be sorted out with the Fibinaccary numbers. The first we discovered above: each integer corresponds to more than one Fibinaccary. Thus $4 = 101.01 = 101.0011$, for instance. In fact, any integer has an infinite number of Fibinaccary representations. From (*) above we see that the rightmost 1 of any Fibinaccary can always be expressed as ...011, and this process can be carried on as we please. This is a serious problem, should we wish to make use of the Fibinaccaries.

Curiously, if we insist that no two adjacent 1s occur, this is a sufficient requirement to ensure uniqueness. We call such an equivalent a *proper Fibinaccary*.

We have made an advance on one front - we could restrict attention to proper Fibinaccaries - but we have a second major problem:

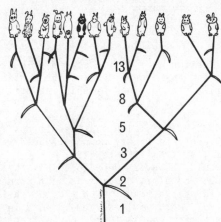

Achillea Ptarmica
(Sneezewort)

or

The Good-natured
Rabbit Breeder
Strikes Again.

M-19

the sequences representing integers are apparently not very different from those that don't represent integers.   For instance, 101.0
is *not* an integer.   How do we tell when a Fibinaccary is an integer?   Eve laChyl is the authoress of an interesting method.

We can look, in fact, at a rather more general problem (the
standard Bourbakiste ploy!): if

$$f(x) = \sum a_n x^n$$

as n ranges through a finite set of integers (positive or negative)
with a_n always an integer, when is $f(\phi)$ an integer?

We define a polynomial, the *Chylean Polynomial*, as follows:

$$C(x) = \sum_{n>0} a_n x^n + \sum_{n<0} a_n (x-1)^{-n}.$$

Since $\phi^{-1} = \phi-1$ it follows that $f(\phi) = C(\phi)$ so we may consider C
instead of f, with the advantage that there are no negative powers.
By the division algorithm there exists a polynomial $q(x)$ and integers a,b such that

$$C(x) = q(x).(x^2-x-1) + ax + b.$$

Then $C(\phi) = q(\phi).0+a\phi+b$ must be an integer, so a = 0 (since ϕ is
irrational).   Thus a necessary and sufficient condition for $f(\phi)$
to be an integer is that when the Chylean polynomial is divided by
x^2-x-1, the remainder should have zero coefficient for x.   And the
value of this integer will be given by b.

Thus, to find out whether $\phi^2+\phi+\phi^{-1}+\phi^{-3}$ is an integer, we form

$$C(x) = x^2 + x + (x-1) + (x-1)^3$$
$$= x^3 - 2x^2 + 5x - 2$$

and divide by x^2-x-1 to get remainder 5x-3.   Since the x-coefficient
is non-zero, the expression is not an integer.

Arithmetic using Fibinaccaries is unfamiliar, but not appreciably
more difficult.   Let us give an example, say of addition.   The
reader can easily check the steps in:

```
    101.01    +                     [4+5 = ?]
   1000.1001
   =========
   1101.1101
  10002.0001
  10010.0101  = φ⁴ + φ + φ⁻² + φ⁻⁴ = 9.
```

$10010.0101 = \phi^4 + \phi + \phi^{-2} + \phi^{-4} = 9.$

Multiplication is a little more difficult, but is aided by the fact
that, as in any binary system, the partial products, being multiples
of 0 or 1, are obtained by shifting strings.

Is there any practical reason why this Fibinaccary system might
be better than the usual binary system?   It is tempting to suggest
not, especially as with the 9 positions required to represent the
digit 9 above, a binary computer could represent all integers from
0 to 511, or from -256 to +255.

It would, however, be a foolish mathematician who would throw
away such an unforseen gift as an alternative number system without
closer investigation.   With the increasing interest being shown in
mechanical logic devices, some points in favour of the Fibinaccary
representation can be found in the fact that during addition the
number of binary 1s around either remains constant or decreases by 1
at each stage.   In contrast, the change from 15 to 16 in regular

binary involves 5 changes of bi-stable units.

But in practice it seems that Fibinaccary is doomed except as a curiosity. The intrigued reader might care to develop his interest in two ways: he might like to prove that given *any* Fibinaccary number - not just an integer - a sufficient condition for uniqueness of representation is that (as before) no two 1s occur together, and also that no infinite sequence 010101010101... appears. Alternatively, he might care to perform similar analyses on other binary systems: known systems exist, based on the binomial coefficients, and also "Grey Code". Par time for the first of these projects should be 15-30 minutes. Are you ready...?

BIBLIOGRAPHY

H.S.M.Coxeter   *Introduction to Geometry* pp. 160-172.
D.E.Knuth   *The art of computer programming* vol. I (1.2.8)
   Addison-Wesley 1968.
R.Saktreger   *Enquiry* (1968) p.39.

* * * * * * * * *

gruppen

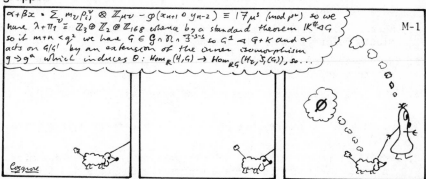

THE COMPUTERIZED ARMY:
MARK I

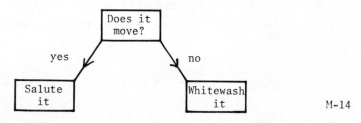

An explorer was captured on a South Sea island by cannibals who confined him to a grass hut while they polished the cooking pot. On consulting his diary, he discovered that at noon the following day there was to be a total eclipse of the Sun. Thinking to impress the natives with his powers (cf. Rider Haggard, Mark Twain) and ensure his release, the explorer consulted the guard-cannibal at the door and asked when he was to be cooked. "At about 11.30," replied the native. "Just before the eclipse."

M-8

SIGNS

25

*In solving the following problems, mathematical know-
ledge will prove less useful than scientific common
sense.   This is not surprising: we stole them from
the Engineering Department's prospectus.   (Let it
never be said that MANIFOLD is narrow and sectarian.)
The answers are given on page 91.*

THINK 13

1. Could you use a suction hovercraft on the ceiling?

2. Why do you put your thumb on a hosepipe to make the water squirt further?

3. Why can't you light a fire with too little fuel, and why do guinea-pigs eat all the time?

4. When a tall brick chimney is 'felled', why does it break in the middle before it hits the ground?

5. Why do American oil-tankers cruise one knot faster than British oil-tankers?

6. Why does a steel tape-measure bend more easily one way than the other?

7. Could a metal cricket bat be as good and effective as a willow one?

8. How is it possible for a yacht to sail upwind?   Could a mirror in space use radiation pressure from the Sun to sail closer to the Sun?

9. Why are cooling towers made the shape they are?

10. Why are most doors hinged rather than sliding?

11. Why don't spanners for small nuts have handles as long as for big nuts?

12. A fuel-tank in a spaceship is partially full.   What happens to fuel under surface tension in the complete absence of gravity? Assume that the fuel wets the walls, i.e. contact angle is zero.

13. A ternary system of numbers employs the basic symbols 1, 0, -1. The decimal number 76 is therefore the ternary number 10-111, and the decimal number -65 is -11-1-11.   What are the advantages of this ternary system over either decimal or binary systems for use in digital computers?

* * * * * * * * *

SIGNS   The paper *Bemerkung über die Einheitengruppen semilokaler Ringe* (Math.-Phys. Semesterber. 17 (1970) 168-181) by Günter Scheja and Uwe Storch is dedicated:
    "To Euclid on his 2300th birthday."   M-11

The validity of a proof is a function of time, no
matter what Bertrand Russell thought.   It is less
a matter of logic; more of conviction.   The
clarity of vision within a new paradigm sometimes
makes previous errors blindingly obvious...

au Courant with Differential Equations

TIM POSTON

Best to reproduce rather than summarize the item from [1], though
many readers will have met it already, as it is so succinctly put
as to defy abbreviation.

> Suppose a train travels from station A to station B along a
> straight section of track.   The journey need not be of uniform
> speed or acceleration.   The train may act in any manner, speed-
> ing up, slowing down, coming to a halt, or even backing up for a
> while, before reaching B.   But the exact motion of the train is
> supposed to be known in advance; that is, the function s = f(t)
> is given, where s is the distance of the train from station A
> and t is the time, measured from the instant of departure.   On
> the floor of one of the cars a rod is pivoted, so that it may
> move without friction either forward or backward until it touch-
> es the floor.   If it does touch the floor, we assume that it
> remains on the floor henceforth; this will be the case if the
> rod does not bounce.   Is it possible to place the rod in such
> a position that, if it is released at the instant when the train
> starts and allowed to move solely under the influence of gravity
> and the motion of the train, it will not fall to the floor
> during the entire journey from A to B?

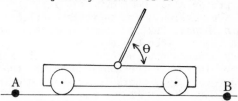

> It might seem quite unlikely that for any given schedule of mot-
> ion the interplay of gravity and reaction forces will always per-
> mit such a maintenance of balance under the single condition that
> the initial position of the rod is suitably chosen.   Yet we
> state that such a position always exists.
>
> Paradoxical as this assertion might seem at first sight, it
> can be proved easily once one concentrates on its essentially
> topological character.   No detailed knowledge of the laws of
> dynamics is needed; only the following simple assumption of a
> physical nature need be granted.   <u>The motion of the rod depends</u>
> <u>continuously on its initial position</u>.   Let us characterise the
> initial position of the rod by the initial angle x which the rod

makes with the floor, and by y the angle which the rod makes
with the floor at the end of the journey, when the train reaches
the point B.   If the rod has fallen to the floor we have either
y = 0 or y = π.   For a given initial position x the end position y is, according to our assumption, uniquely determined as a
function y = g(x) which is continuous and has the values y = 0
for x = 0 and y = π for x = π (the latter assertion simply exp-
ressing that the rod will remain flat on the floor if it starts
in this position).   Now we recall that g(x), as a continuous
function in the interval 0 ≤ x ≤ π, assumes all the values between g(0) = 0 and g(π) = π; consequently, for any such value y,
in particular y = π/2, there exists a specific value of x such
that g(x) = y; so there exists an initial position for which the
end position of the rod is perpendicular to the floor.   (Note:
in this argument it should not be forgotten that the motion of
the train is fixed once for all.)
     Of course the reasoning is entirely theoretical.   If the
journey is of long duration or if the train schedule, expressed
by s = f(t), is very erratic, then the range of initial positions x for which the end position g(x) differs from 0 or π will
be exceedingly small, as is known to anyone who has tried to
balance a needle upright on a plate for an appreciable time.
Still, our reasoning should be of value even to a practical mind
inasmuch as it shows how qualitative results in dynamics may be
obtained by simple arguments without technical manipulation.

Unfortunately, it is wrong.   Not mathematically, but in assuming
continuity for this problem.   Within its spirit, let us neglect
the probabilistic aspect brought in by, e.g., any realistic model
of friction effects.   Stick with the assumption that the motion of
the rod when not on the floor is given by a differential equation
D (time-dependent thanks to the motion of the train).   A nice enough D - say with differentiable coefficients - to guarantee that
every initial position x determines a unique y = g(x) on arrival at
B.   (For examples without unique solutions, see p.38 of [2], which
is as cheery and pictorial as MANIFOLD itself.)
     Look first at the version where there is no floor: just a pivoted
lever free to turn through 360°.   The initial angle can be any x
between -180° and 180°.   Then D's niceness lets us *prove* the standard result that y depends continuously on x.   This implies, using
just a bit more topology that the above appeal to the Intermediate
Value Theorem, that for any y there is at least one initial angle x
with y = g(x).
     Fig.1 (opposite) shows a typical set of histories, for a general
D and a rather short trip.   Notice that sometimes *more* than one x
gives arrival at the same y - with, of course, different velocities.
(Given 'final conditions' y and angular velocity $\dot\theta_y$, if D is second
order we can compute back to find *unique* x and $\dot\theta_x$ that would end up
at $(y,\dot\theta_y)$.   But the 'final condition' y without $\dot\theta_y$ is not enough,
even if we stipulate $\dot\theta_x$ = 0 as in the figure.)
     The way this plurality begins should be ringing bells in readers
of MANIFOLD 14 and 15: it is one of the many ways catastrophe theory
comes into entirely *non*-'gradient dynamics' situations.   That's why

28

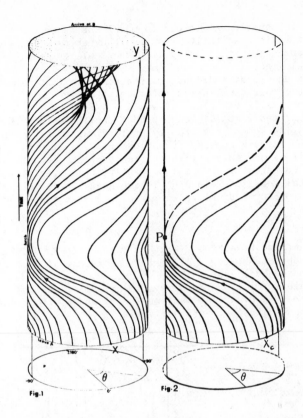

Fig.1 Fig.2

it's shown: it has nothing to do with the main point of this note,
which is:-

Now put those 'absorbing boundary conditions' back - the floor
it stays on if it hits. Then for the same time dependent dynamic D
in the region $-90° < x < 90°$, the picture reduces to Fig.2. So
for the original problem there is a critical angle x_c, and g takes
x to

$$y = \begin{cases} -90° & \text{if } x \leq x_c \\ 90° & \text{if } x > x_c, \end{cases}$$

which is *not* continuous. There is *no* x for which the rod never
falls.

Of course additional hypotheses on D can rescue the original conclusion. For example, one can forbid points like P in Fig.2. But
to *check* that D gives no such points, one very definitely needs some
'detailed knowledge of the laws of dynamics'. Given the usual laws,
the only physical assumptions I can find which guarantee a nonfal-
ling history are that the pivot is perfect with the movement of the
train perfectly, totally horizontal. (Not just level track: the
train must have no springs. Why?)

In some modern uses of differential equations the hypothesis 'no
points like P' (technically the *isolating block condition*) can be
proved under much more reasonable assumptions. For example, it is
a powerful tool in proving the existence of nerve-impulse-like

29

solutions to equations intended to model the behaviour of nerves
[3].   But valid qualitative results in dynamics often need not-so-
simple arguments, with topological technicalities replacing the
traditional kind.

Courant and Robbins did not make a *silly* mistake - but Dynamical
Systems has progressed in the intervening 1/3 century so as to make
their error easily visible - it *would* be silly *now*.   (On the other
hand Figs. 1 and 2 would probably have convinced them they were
wrong; imagine going back and trying to dissuade say, Kant of any
opinion now largely opposed by philosophers.   If philosophy progresses, it is not in a way that makes *anything* clearly wrong to
everybody competent.)   More importantly, this progress is reaching
the original goals of Poincaré: I have heard 'practical minds' coming away from topological talks marvelling at the power of the methods used.

BIBLIOGRAPHY

1. R.Courant and H.Robbins   *What is Mathematics?* O.U.P. 1941.
2. R.L.E.Schwarzenberger   *Elementary Differential Equations*,
   Chapman and Hall 1969.
3. C.C.Conley   *On travelling wave solutions of nonlinear diffusion
   equations*, in Dynamical Systems Theory and Applications, Lecture
   Notes in Physics 38, Springer Verlag 1975.

gruppen   * * * * * * * * *

THE COMPUTERIZED ARMY: M-2
MARK II

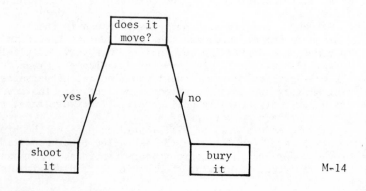

Archimedes, so the legend goes, got his inspiration
in the bath.   Henry Squarepoint got his from the
bath - the taps, in fact.   Another impressive example
of the power of Catastrophe Theory to shed light on
the most varied and mundane topics.   But the real
reason for this article is that it is a puzzle: WHO IS
HENRY SQUAREPOINT? *

my bath-tap is a cusp catastrophe

HENRY SQUAREPOINT

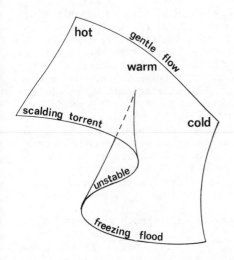

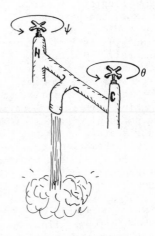

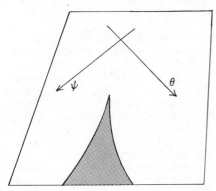

Total flow is a nice convex
function of (θ,ψ); opening one
tap further always gives more
water.   But *temperature* isn't
a function of (θ,ψ), i.e.
single-valued, at all: it's a
catastrophe.

The warmth of gentle flow
is smoothly adjustable, but a
large flow either freezes or
scalds.   The middle range is
inaccessible.

CURSES!

*

WHO IS JERRY CORNELIUS?

HENRY SQUAREPOINT is Henri Poincaré: "Point" = point; "carré" =
square.   But actually, he's Jerry Cornelius.   The question is,

31

*There really ought to be a Nobel Prize for Mathematics.
The case is far stronger than for, say, economics - and
that has a Nobel only because it bought itself one, by
way of the banking world. Even as we write, the theories
of one Nobel-winning economist are being tested to dest-
ruction, along with the British economy. But Mathematics
has an equally prestigious award, even though, in terms of
cash value or publicity, it is more modest. It is:*

the Fields medal

DANIEL DROLET

When in December of every year, the world's physicists, chemists,
physiologists, writers, and peace-makers are honoured with Nobel
prizes, the mathematicians of the world feel a little left out.
One would think that the great man Nobel would have seen fit to include in his list a category for fine work in mathematics. Unfortunately, he didn't. Why, remains a mystery. There are rumours
of sharp discord between Nobel and the mathematician Mittag-Leffler,
even talk of a romantic triangle - although there is no evidence
for this, or even that either man was ever married.

Fine work in mathematics deserves world recognition; and the International Congress of Mathematicians makes such an award every
four years when it meets in convention. A surprising feature of
this mathematical equivalent of the Nobel Prize is that it was the
idea of a Canadian mathematician, it is (in universal colloquial
usage) named after him, the prize money is in Canadian dollars,
and the medal is minted in Ottawa.

The award is known as the Fields Medal - though it has no offi-
cial name and Fields's name does not appear anywhere on it. It
takes this name from the late J.C.Fields (Toronto). It is fairly
well known within mathematics, but almost unknown to the 'outside
world' - and even mathematicians tend to know very little about J.C.
Fields.

John Charles Fields was born in Hamilton, Ontario, on 14th May
1863. He graduated from the University of Toronto in 1884 with a
BA in mathematics, obtained a Ph.D. from Johns Hopkins University
(Baltimore) three years later, and taught in the USA before travelling to Europe to study. In 1902 he was back at Toronto as a lec-
turer. He remained there until his death 30 years later, as Asso-
ciate Professor, Professor, and finally Research Professor.

He seems to have been a man-about-the-world: he was involved
with the Royal Societies of both London and Canada, the Coimbra Institute of Portugal, the Russian Academy of Sciences, the Royal Canadian Institute, the International Mathematical Union, and both
the British and American Associations for the Advancement of Science. He knew many famous mathematicians, including an enduring
friendship with... Mittag-Leffler.

In 1924 he persuaded the ICM to meet in Toronto. This was regarded as 'acceptable' because it was in North America, away from a
Europe fresh from the 1st World War. To promote the conference he
travelled extensively and, thanks to his organizing abilities, the

funds provided by the Ontario and Dominion governments proved more than sufficient.

Shortly before his death on August 9th 1932 he had been promoting a proposal to the ICM to create an award using these surplus funds. This was accepted by the 1932 Congress, and the first Fields Medals presented in Oslo in 1936. Originally two medals were given at each ICM - every 4 years - but since 1966 the number has doubled to four each meeting (except 1974). There are two basic criteria for Fields medallists: they must be young (generally agreed to mean under 40), and they must have made outstanding contributions to mathematics.

THE MEDAL

The medal is of gold, 2.5 inches in diameter, and was designed by the distinguished Canadian sculptor R. Tait McKenzie. On the obverse is the head of Archimedes, ARCHIMEDOUS in Greek capitals, the artist's monogram RTM, and the date MCMXXXIII. The inscription 'transire suum pectus mundoque potiri' freely translates as 'to transcend one's human limitations and master the universe', a quotation from the Roman poet Manilius. The reverse bears the inscription 'Congregati ex toto orbe mathematici ob scripta insignia tribuere': 'mathematicians gathered together from the whole world honour noteworthy contributions to knowledge'. Archimedes' diagram of the sphere and circumscribed cylinder is discernible in the background. The name of the medallist is engraved along the edge of the medal.

The medal is accompanied by a modest cash prize.

The article above first appeared in Carleton Coordinates *magazine.*
The cartoon below first appears in Seven Years of MANIFOLD 1968-80.

Every so often, MANIFOLD ventured into what can only be described as Mathematical Theology. In this article, a worker at the Institute for Unsound and Vibration Research (Southampton) offers a new use for topological dynamics:

Sanctification and the Hopf Bifurcation

PHILIP HOLMES

Recent speculations on biological oscillations [1] and in particular models of brain behaviour involving oscillations [2] have encouraged the author to reveal a sample of his own work. For some years he has been interested in the halo as a symbol of holiness and sanctification: an aura of (intermittently visible)energy surrounding the heads of Saints (Fig. 1). The manner in which the halo develops is especially interesting, since there appear to be few records of children born with haloes, excepting, naturally, the Christ-child (clearly a non-generic case). The reasoning is thus: in seeking a simple model of mental states, represent the brain of the average man as a *sink* p_S in some *ambient mental state-space* M, since it acts as recipient of innumerable bits of advice, facts, theorems, etc. (Ignore for the moment the case of the obstinate, forgetful fellow, clearly a saddle-point $p_C \in M$.) The mind of a holy man, however, is more likely to be a *source* $p_u \in M$. Now let some flow $\phi_t : M \to M$, parametrised by time t, represent the transfer of mental energy (ideas). The problem then reduces to that of describing the *bifurcation* of the fixed point $p_S \to p_u$ as the potential holy man 'lights up'. The halo provides the obvious clue: an *attracting closed orbit* γ is thrown off by the Saint's mind, which itself becomes a source. We therefore have:

PROPOSITION 1: Sanctification generically consists in the creation of an attracting closed orbit in a Hopf bifurcation [3].

Fig.1 St. Sebald on the Column, by Dürer, c. 1501. Note vague appearance of attracting orbit and unusual radial flowlines.

Certain works indicate the existence of multiple haloes (Fig. 2), which can be modelled by successive Hopf bifurcations, alternating attracting and repelling orbits being created (cf. R.Thom's model of the evolution of

Fig. 2
St. Sebald in the Niche,
Dürer, 1518.

a nebula in the material universe, [1] chapter 6.   Takens has recently published work on generalized Hopf bifurcations [4] but the present major application has clearly escaped him.   In this respect the wide halo of Fig. 1 might be seen as a structurally unstable 'vague attractor' which subsequently bifurcates to the structurally stable triple orbit of Fig.2.   Could this be an early example of the *generic evolution* of an attractor?

The validity of the theory sketched here could only be tested by a massive survey of the European artistic heritage, with a view to investigating the growth in size, with time, of the closed orbits (haloes) of given Saints.   Hopf's Theorem predicts a growth rate proportional to $\sqrt{\varepsilon}$ where ε = time elapsed after bifurcation (sanctification) [3].   The author has already contacted theological and artistic experts and hopes to persuade the appropriate bodies to finance an initial sabbatical year in Italy in order to place this study on a firm basis.

The author would like to thank his colleagues for their infinite patience, gentle care, and encouraging suggestions.

BIBLIOGRAPHY

1. R. Thom   *Stabilité Structurelle et Morphogénèse* 1972.
2. E.C.Zeeman   ICTP Conf. on Neural Networks, Trieste 1972.
3. E.Hopf   *Ber. Math. Phys. Kl. Sächs. Akad. Wiss.* Leipzig 94 (1942) 1-22.
4. F.Takens   *J. Diff. Eq.* 14 (1973) 476-493.

*This, O Best Beloved, is another tale of the High
and the Far-Off Times. In the blistering midst
of the Sand-Swept Sahara lived a Pride of Lions.
There was a Real Lion, and a Projective Lion, and
a pair of Parallel Lions; and all manner of Lion
Segments. And on the edge of the Sand-Swept
Sahara there lived a 'nexorable Lion Hunter...*

15 new ways to catch a Lion

JOHN BARRINGTON

I make no apologies for raising once again the problems of the mathematical theory of big game hunting. As with any branch of mathematics, much progress has been made in the last decade.

The subject started in 1938 with the epic paper of Pétard [1].
The main problem is usually formulated as follows: *In the Sahara
desert there exist lions. Devise methods for capturing them.*
Pétard found ten mathematical solutions, which we can paraphrase
as follows.

1. *The Hilbert Method*. Place a locked cage in the desert. Set
up the following axiomatic system.

(i) The set of lions is non-empty.

(ii) If there is a lion in the desert, then there is a lion in
the cage.

Theorem 1: There is a lion in the cage.

2. *The Method of Inversive Geometry*. Place a locked, spherical
cage in the desert, empty of lions, and enter it. Invert with
respect to the cage. This maps the lion to the interior of the
cage, and you outside it.

3. *The Projective Geometry Method*. The desert is a plane. Project this to a line, then project the line to a point inside the
cage. The lion goes to the same point.

4. *The Bolzano-Weierstrass Method*. Bisect the desert by a line
running N-S. The lion is in one half. Bisect this half by a
line running E-W. The lion is in one half. Continue the process
indefinitely, at each stage building a fence. The lion is enclos-
ed by a fence of arbitrarily small length.

5. *The General Topology Method*. Observe that the desert is a sep-
arable metric space, so has a countable dense subset. Some sequence converges to the lion. Approach stealthily along it, bearing
suitable equipment.

6. *The Peano Method*. There exists a space-filling curve passing
through every point of the desert. It has been remarked [2] that
such a curve may be traversed in as short a time as we please. Ar-
med with a spear, traverse the curve faster than the lion can move
his own length.

7. *A Topological Method*. The lion has at least the connectivity
of a torus. Transport the desert into 4-space. It can now be
deformed in such a way as to knot the lion [3]. He is now help-
less.

8. *The Cauchy Method*. Let f(z) be an analytic lion-valued

function, with ζ the cage.   Consider the integral

$$\frac{1}{2\pi i} \int_C \frac{f(z)}{z - \zeta} \, dz$$

where C is the boundary of the desert.   Its value is $f(\zeta)$, that is, a lion in a cage.

9. *The Wiener Tauberian Method.*   Procure a tame lion L_0 of class L($-\infty,\infty$) whose Fourier transform [Furrier transform?] nowhere vanishes, and set it loose in the desert.   Being tame, it will converge to the cage.   By Wiener [4] every other lion will converge to the same cage.

10. *The Eratosthenian Method.*   Enumerate all objects in the desert; examine them one by one; discard all those that are not lions.   A refinement will capture only prime lions.

Pétard also gives one physical method with strong mathematical content:

11. *The Schrödinger Method.*   At any instant there is a non-zero probability that a lion is in the cage.   Wait.

The next work of any significance is that of Morphy [5].   I confess that I do not find all of his methods convincing.   The best are:

12. *Surgery.*   The lion is an orientable 3-manifold with boundary and so [6] may be rendered contractible by surgery.   Contract him to Barnum and Bailey.

13. *The Cobordism Method.*   For the same reasons the lion is a handlebody.   A lion that can be handled is trivial to capture.

14. *The Sheaf-theoretic Method.*   The lion is a cross-section [8] of the sheaf of germs of lions in the desert.   Re-topologize the desert to make it discrete: the stalks of the sheaf fall apart and release the germs, which kill the lion.

15. *The Postnikov Method.*   The lion, being hairy, may be regarded as a fibre space.   Construct a Postnikov decomposition [9].   A decomposed lion must, of course, be long dead.

16. *The Universal Covering.*   Cover the lion by his simply-connected covering space.   Since this has no holes, he is trapped!

17. *The Game-theory Method.*   The lion is big game, hence certainly a game.   There exists an optimal strategy.   Follow it.

18. *The Feit-Thompson method.*   If necessary add a lion to make the total odd.   This renders the problem soluble [10].

Recent, hitherto unpublished, work has revealed a range of new methods:

19. *The Field-theory Method.*   Irrigate the desert and plant grass so that it becomes a field.   A zero lion is trivial to capture, so we may assume the lion $L \neq 0$.   The element 1 may be located just to the right of 0 in the prime subfield.   Prize it apart into LL^{-1} and discard L^{-1}.   (Remark: the Greeks used the convention that the product of two lions is a rectangle, not a lion; the product of 3 lions is a solid, and so on.   It follows that every lion is transcendental.   Modern mathematics permits algebraic lions.)

20. *The Kittygory Method.*   Form the category whose objects are the lions in the desert, with trivial morphisms.   This is a small category (even if lions are big cats) and so can be embedded in a concrete category [11].   There is a forgetful functor from this to the category of sets: this sets the concrete and traps the embedded lions.

21. *Backward Induction.*   We prove by backward induction the statement L(n): "It is possible to capture n lions".   This is true for sufficiently large n since the lions will be packed like sardines and have no room to escape.   But trivially L(n+1) implies L(n) since, having captured n+1 lions, we can release one.   Hence L(1) is true.

22. *Another Topological Method.*   Give the desert the *leonine* topology, in which a subset is closed if it is the whole desert, or contains no lions.   The set of lions is now dense.   Put an *open* cage in the desert.   By density it contains a lion.   Shut it quickly!

23. *The Moore-Smith Method.*   Like (5) above, but this applies to non-separable deserts: the lion is caught not by a sequence, but by a net.

24. *For those who insist on sequences.*   The real lion is non-compact and so contains non-convergent subsequences.   To overcome this let Ω be the first uncountable ordinal and insert a copy of the given lion between α and $\alpha+1$ for all ordinals $\alpha < \Omega$.   You now have a *long lion* in which all sequences converge [12].   Proceed as in (5).

25. *The Group Ring Method.*   Let Γ be the free group on the set G of lions, and let $\mathbb{Z}\Gamma$ be its group ring.   The lions now belong to a ring, so are circus lions, hence tame.

26. *The Bourbaki Method.*   The capture of a lion in a desert is a special case of a far more general problem.   Formulate this problem and find necessary and sufficient conditions for its solution. The capture of a lion is now a trivial corollary of the general theory, which *on no account should be written down explicitly.*

27. *The Hasse-Minkowski Method.*   Consider the lion-catching problem modulo p for all primes p.   There being only finitely many possibilities, this can be solved.   Hence the original problem can be solved [13].

28. *The PL Method.*   The lion is a 3-manifold with non-empty boundary.   Triangulate it to get a PL manifold.   This can be collared [14], which is what we wish to achieve.

29. *The Singularity Method.*   Consider a lion in the plane.   If it is a regular lion its regular habits render it easy to catch (e.g. dig a pit).   WLOG it is a singular lion.   Stable singularities are dense, so WLOG the lion is stable.   The singularity is not a self-intersection (since a self-intersecting lion is absurd) so it must be a cusp.   Complexify and intersect with a sphere to get a trefoil knot.   As in (7) the problem becomes trivial.

30. *The Measure-Theoretic Method.*   Assume for a contradiction that no lion can be captured.   Since capturable lions are imaginary, all lions are real.   On any real lion there exists a non-trivial invariant measure μ, namely Haar or Lebesgue measure.   Then $\mu \times \mu$ is a Baire measure on L×L by [15].   Since a product of lions cannot be a bear, the Baire measure on L×L is zero.   Hence $\mu = 0$, a contradiction.   Thus all lions may be captured.

31. *The Method of Parallels.*   Select a point in the desert and introduce a tame lion not passing through that point.   There are three cases:

  (a) The geometry is Euclidean.   There is then a unique parallel lion passing through the selected point.   Grab it as it passes.

  (b) The geometry is hyperbolic.   The same method will now catch infinitely many lions.

(c) The geometry is elliptic. There are no parallel lions, so every lion meets every other lion. Follow a tame lion and catch all the lions it meets: in this way every lion in the desert will be captured.

32. *The Thom-Zeeman Method.* A lion loose in the desert is an obvious catastrophe [16]. It has three dimensions of control (2 for position, 1 for time) and one dimension of behaviour (being parametrized by a lion). Hence by Thom's Classification Theorem it is a swallowtail. A lion that has swallowed its tail is in no state to avoid capture.

33. *The Australian Method.* Lions are very varied creatures, so there is a variety of lions in the desert. This variety contains free lions [17] which satisfy no non-trivial identities. Select a lion and register it as "Fred Lion" at the local Register Office: it now has a non-trivial identity, hence cannot be free. If it is not free it must be captive. (If "Fred Lion" is thought to be a trivial identity, call it "Albert Einstein".)

BIBLIOGRAPHY

1. H.Pétard   *A contribution to the mathematical theory of big game hunting*, Amer. Math. Monthly 45 (1938).
2. E.W.Hobson   *The theory of functions of a real variable and the theory of Fourier's series*, 1927.
3. H.Seifert and W.Threlfall   *Lehrbuch der Topologie*, 1934.
4. N.Wiener   *The Fourier integral and certain of its applications*, 1933.
5. O.Morphy   *Some modern mathematical methods in the theory of lion hunting*, Amer. Math. Monthly 75 (1968) 185-7.
6. M.Kervaire and J.Milnor   *Groups of homotopy spheres I*, Ann. of Math. 1963.
7. This footnote has been censored by the authorities.
8. It has been verified experimentally that lions are cross.
9. E.Spanier   *Algebraic Topology*   McGraw Hill 1966.
10. W.Feit and J.G.Thompson   *Solvability of groups of odd order*, Pac. J. Math. 1963.
11. P. Freyd   *Abelian Categories*.
12. J.L.Kelley   *General Topology*.
13. J.Milnor and D.Husemoller   *Symmetric Bilinear Forms* 1973.
14. C.P.Rourke and B.L.Sanderson   *Introduction to Piecewise Linear Topology* 1973.
15. S.K.Berberian   *Topological Groups*.
16. R.Thom   *Stabilité Structurelle et Morphogénèse* 1972.
17. Hanna Neumann   *Varieties of Groups* 1972.

M-6

If God had wanted us to write haiku,
He would have given us 17 fingers...

According to The Hitch Hiker's Guide to the Galaxy, M-3
the Answer to the Great Question of Life, the
Universe, and Everything, is 42. Deep Thought
wasn't far off: the True Answer is 49.

a small world

BARRY PILTON

We are going to create a universe.

Commensurate with our powers as minor deities, it will be a small universe; but we shall populate it with particles capable of moving about, coalescing, or splitting into several parts. We shall study our universe, and from our deductions we shall draw several morals for the would-be experimenter or philosopher.

To business, then.

To confine our particles to a small space we take a hint from plasma physics, and make our universe toroidal in shape. In fact we take the set of all ordered pairs (x,y) where x and y are integers modulo 7. We can picture this as a 7×7 rectangle, with opposite edges identified in the usual manner. There are 49 points in our universe.

This has quantized space - so we will quantize time. Time, indicated by t, will take only positive integer values.

Finally, our particles. If P is a particle, we endow it with position
$$(x^P(t), y^P(t))$$
and mass
$$m_P(t).$$

The mass will also be a positive integer.

Any set of particles which is connected (in the sense that points next to each other in horizontal or vertical directions are connected, but not diagonal ones) will be said to form a *molecule*. Thus

or are molecules, but

is not.

Heavy particles move slowly; so a particle of mass m will only move at time t if m divides t. The Law of Motion will be
$$x^P(t) = \sum x^Q(t-1) + 1$$
$$y^P(t) = \sum y^Q(t-1) + 1 \qquad (*)$$
where the summations are over all particles Q not in the same molecule as P. (The empty sum, of course, is zero.)

If at time t two particles of masses m, n are both in the same place, they coalesce to form a single particle of mass m+n.

After particles have moved and coalesced, they may also decay. A

40

particle of mass m at (x,y) will decay if and only if
  (1) t is a multiple of m,
  (2) x or y ≡ t (mod 7).
If x ≡ t (mod 7), it decays into two particles, each of mass m, at
points (x±1,y).   If y ≡ t it decays into two particles of mass m
at (x,y±1). If both x and y ≡ t, it does both, becoming four particles at (x±1,y±1).

  Thus a particle of mass 5 at (1,3) at time 15 will decay into
a particle of mass 5 at (0,3) and another at (2,3).

  Denoting a particle of mass m at (x,y) by xy^m we can write this
symbolically as

$$\begin{array}{ccc} & 13^5 & \\ & \diagup \quad \diagdown & \\ 15 \qquad 03^5 & & 23^5. \end{array}$$

We now start the universe at t = 0 with a particle of mass 1 at
(3,2), and see what happens.

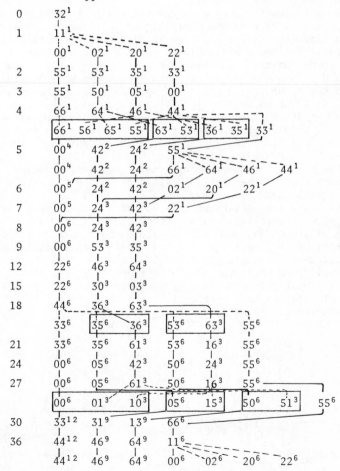

(Here <u>dotted lines</u> are decays, and ⊡boxes⊡ indicate molecules.)

41

A hypothetical scientist or philosopher observing this behaviour would probably notice two things:

The Law of Symmetry: for $t > 0$, if there is a particle xy^m there is also a particle yx^m.

The Law of Increasing Mass: the total mass of the universe is an increasing function of time.

It is easy to prove these laws from the full laws of motion (*). But observe the danger which the Law of Symmetry may lead the unwary cosmologist into: the line of reasoning "The universe is symmetric, and as far as our observations go, it has always been symmetric. Therefore, in a "Little Bang" theory of creation, the original particle must have been symmetrically placed, at some point (x,x)."

But actually, it wasn't, it was at $(3,2)$. It then went to $(1,1)$; and it would have done the same from any other starting position. Symmetry need not run backwards, and maybe the Steady State Theory holds after all.

There is something else rather curious. From time 8 onwards, all particles have masses divisible by 3. From the laws of motion it is clear that if at any time the 'species' of particles of mass divisible by some d has achieved total dominance, then it will be self-perpetuating. In this case a self-perpetuating species has evolved with d = 3. The point is that the number 3 is unexpected: it is not clear that 3 plays any special role in the laws of motion. It turns up more by luck than judgement. *But once it has turned up it won't go away again.* If conditions permit the existence of a self-perpetuating species (life) then it is not surprising if such lifeforms appear accidentally, nor is it surprising if they take an unpredictable form.

This particular lifeform evolves rather successfully. By time 48 every particle has mass divisible by 9, and the species has evolved to a higher level.

The basic laws governing this universe are extremely simple, and the universe itself is small. Despite this, it displays a large range of phenomena, some easy to see, others more mysterious. For example, up to time 54, only 27 of the 49 points have at any time been occupied by particles. Why?

Complicated phenomena may be governed by simple laws. But to discover what those laws are by analysing observations, even in the present case, is very difficult.

Experimental scientists take heed!

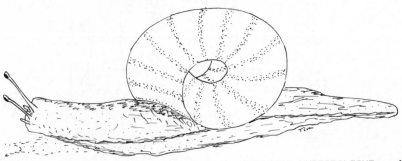

ESCARGOT HYPERBOLIQUE   M-19

Look, let's put it this way: Catastrophe Theory
may not be the Greatest Thing Since Sliced Bread,
but it's not the Bermuda Triangle either.   The
real question is: how did René Thom think of it
to begin with?   Here's MANIFOLD's theory...

Portrait of René Thom

THEODOR BRÖCKER

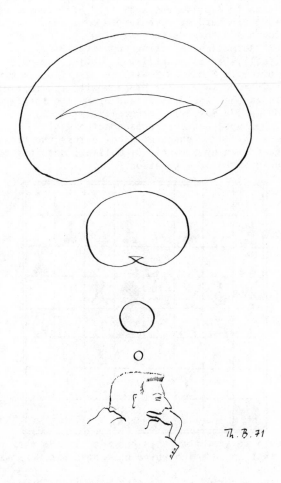

"I would draw your attention, Watson, to the strange affair of the bishop on square f1."
"But there is no bishop on f1, Holmes."
"That," remarked Holmes drily, "is what is so strange."

the one-move mate

STANLEY COLLINGS

The following chess problem is particularly delightful, not least because every piece makes a contribution to the solution. It is not our intention to deprive the dedicated *afficianado* of the pleasure of puzzling his head at some length over this one, so we present only a few hints towards the solution, arranged progressively throughout this selection, starting opposite.

"You may recall, Watson, that I was foolish enough to allow you to rush into print in MANIFOLD-2 with an assertion of mine that the only respectable move in a chess problem with the instructions 'move and mate in one' would be a pawn promotion to knight?"

"Indeed, Holmes, a masterly stroke if ever I..."

"Masterly, perhaps, but erroneous nonetheless!"

Holmes passed a flimsy sheet of paper across the breakfast table to me. "A gentleman by the name of Collings has just sent this to me. What do you make of it, Watson?"

"You may care to know that this position is one that might arise in play, and that in addition white has moved his QRP twice during the last ten moves..."

There elapsed a period of some minutes while I studied the scrap of paper.

44

"There are no mating moves, Holmes!" I was forced to cry.
"Come now Watson!   I expected better of you than that.   Perhaps you have not considered:
         P x P e.p. *mate* (a5 → b6)."
"Ah yes, Holmes - an *en passant* capture.   I had seen that possibility, but rejected it - we have no way of knowing that Black's last move was
         .....   P - QKt4."
"Except that White *does* mate in a single move, and once we have eliminated the impossible, all that remains, however improbable... But I expect you yourself are capable of completing that epigram, Watson."
"Simple then, Holmes!   There is no other possible move, therefore White takes en passant, and wins -"
"Unless, of course, Watson," said Holmes with a touch of a smile flickering about his face, "White castled on the King-side..."

The next hint is on page 58

* * * * * * * * *

It was Leopold Kronecker who insisted on dragging   M-6
religion into Set Theory.   Blame him, then, for:

ontology revisited

VOX FISHER

THEOREM (due to Anselm, Aquinas, and others.)

The Axiom of Choice is equivalent to the existence of a unique God.

PROOF:

$\Longrightarrow$: (Assuming the equivalence of the Axiom of Choice and Zorn's Lemma)˙Partially order the set of subsets of the set of all properties of objects by inclusion.   This set has maximal elements.   God is by definition (Anselm) one of these maximal elements.   Now
    God $\subseteq$ God $\cup$ {existence}, so God = God $\cup$ {existence}
Therefore God exists.
    To prove uniqueness, let God and God' be two gods; then God $\cup$ God' $\subseteq$ God (Aquinas), therefore God $\supseteq$ God'; similarly God' $\supseteq$ God; hence God = God'.

$\Longleftarrow$: Given a set $\{A_\alpha\}_{\alpha \in A}$ of sets, let the unique God pick $x_\alpha \in A_\alpha$ for each $\alpha \in A$.   (He can do so by omnipotence, proved as for existence above.)   Then $(x_\alpha)_{\alpha \in A} \in \Pi_{\alpha \in A} A_\alpha$ as required.

Throughout the 1960's and 70's, prodigious effort
has been expended on the classification of finite
simple groups.   The solution, now close at hand,
occupies some 20,000 pages.   Here's how it
looked in 1968...

Simple Groups

EVE LaCHYL

Over the past decade, abstract algebra has suffered a minor population explosion all its own.   The object which has been propagating
itself with such overwhelming success is known to the algebraist
as a *simple group*.   For anyone not familiar with this terminology,
suffice it to say that a group is an abstract algebraic structure,
and a simple group is one which cannot be broken down into smaller
groups: it is *atomic*.   More technically, a group is simple if it
has no non-trivial normal subgroups.

A major unsolved problem of group theory has been the classification of all finite simple groups.   To *classify* a class of things
in mathematics one must first find as many of them as one can, and
then prove that no more of them remain to be found.   (Thus, to
classify the even prime numbers one first finds the number 2, then
proves that no other even number is prime.)   Such a classification
would be important because of the atomic nature of simple groups,
but it would not lead automatically to the solution of all the problems of group theory.   Even if the atoms are understood, the
'compounds' need investigating.   But it must be admitted that the
reason why simple groups are receiving so much attention at the moment is partly their importance, partly the lure of a notorious unsolved problem, and partly just a matter of fashion - which has a
greater hold on mathematics than is often admitted.

The most trivial type of simple group is the cyclic group of
prime order.   Not only does this have no non-trivial *normal* subgroups - it has no non-trivial subgroups of any kind, normal or
not!   Less trivial are the *alternating groups* A_n where $n > 4$,
found by Galois.   The simplicity of A_5 was the final step in Galois's proof that the general quintic equation

$$ax^5+bx^4+cx^3+dx^2+ex+f = 0$$

cannot be solved by repeated extraction of radicals.   These groups
are examples of a type of simple group that comes in an *infinite
family*: a whole collection of simple groups constructed in a similar fashion.   The other type is the *sporadic* simple group, which
does not seem to belong to any family.   The first examples of these
were found by Mathieu in 1861 and 1873, and are commonly designated
M_{11}, M_{12}, M_{22}, M_{23}, and M_{24}.   The 'sporadicity' of any particular
simple group may be an inherent property, or it may merely reflect
our ignorance - we have as yet failed to find other members of the
family.

46

Between 1954 and 1968 the number of infinite families known doubled from 9 to 18. Between 1964 and 1968 the number of sporadic groups increased from 5 to 14. Reluctantly, we will leave the infinite families to their own devices (save to remark that the names to look out for are Chevalley, Steinberg, Suzuki, and Ree). Instead, we shall discuss the sporadic simple groups, which comprise the current area of activity.

Apart from the Mathieu groups, no other sporadic simple groups were known until Zvonimir Janko discovered one in Australia (hopping about in the bush, no doubt) in 1965. He found it while trying to characterize the simple groups of a particular special type, and discovered that as well as those already known there was this extra one - call it J_1. In 1967 He published evidence that two further groups, J_2 and J_3, might also exist: in fact he constructed a *character table*. Every group has a character table, but not everything that looks like a character table belongs to a group. Thus it came as no surprise when Graham Higman (Oxford) received a letter from Walter Feit (Yale) showing that J_2 did not exist. Except that one the same day he got one from Marshall Hall proving that it *did* exist. The fundamental flaw in mathematics? Not this time. Janko had made a mistake in the character table, which is what Feit had found. Nevertheless, there *was* a group. Marshall Hall's existence proof was carried out with the help of the computer TITAN. A computer again played a decisive role when in 1968 Higman, John MacKay, and ATLAS proved that J_3 also existed. Higman received the news just before giving a lecture titled *Does the Big Janko Group exist?* and started the lecture by writing the word 'YES' on the blackboard, adding: 'Does anyone want to know any more?'

That made two groups for 1967, but more were to come. Donald Higman (≠ Graham) and Sims stuck a lot of Mathieu groups together, and found a new simple group. Graham Higman found one of the same order... with the same character table... and indeed it turned out to be the same group, looked at in a different way. Suzuki found another, using a geometrical method.

1968 began as 1967 left off, when MacLaughlin found yet another simple group, again geometrically. Geometry appeared to be creepin into the subject, even though it started off as algebra. And geometry was to play an unexpected part in the most sensational development so far. Up till now, no more than two new groups had been found at once (and those by Janko). But between the first and final drafts of this article, John Conway (Cambridge) found *three* - it may be more by now, the details are still being worked out - the largest of which contains *all but two* of the known sporadic simple groups.

The story begins in a totally unrelated part of mathematics: sphere-packing. The basic problem here is how to pack large numbers of equal spheres in the most economical manner. In two dimensions, the answer is the packing on the right. But in higher dimensions, less is known. John Leech discovered a very economical packing in 24 dimensions. He

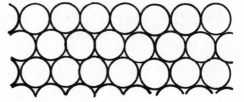

suspected that the symmetry group of this packing might have interesting properties, and tried to 'sell' the group to a number of mathematicians, meeting much sales resistance, until Conway became interested.   After a lot of hard work, Conway found that the group had order $2^{22}.3^9.5^4.7^2.11.13.23$, and telephoned John Thompson, one of the world's foremost group-theorists, who was in Cambridge at the time.   'I've got a subgroup of $O_{24}(Z)$ of order...' said Conway. Ten minutes later Thompson rang back.   'It's got a simple quotient of half the order.'

Conway set about discovering more properties of the group, and in the process found two more simple groups as subgroups.   There may well be more lurking inside it, as yet undiscovered.

In a subject which starts as algebra and ends up involving 24-dimensional sphere-packings, it is hard to predict what will happen next.   Perhaps the solution to the classification problem is just around the corner.   Perhaps the problem has no sensible solution. All that seems certain is that future developments are likely to be as peculiar as those of the past.

BIBLIOGRAPHY

R.W.Carter   *Simple Groups and Simple Lie Algebras*, Journal of the London Mathematical Society 40 (1965) 193-240.

* * * * * * * * *

That was 1968.   By 1981 so much more has happened that we shall provide a brief:

Update

The number of sporadic simple groups is now 26; there are no new infinite families; *the list is complete*.   According to Michael Aschbacher and Daniel Gorenstein, prime movers of a programme to classify all finite simple groups, recent results of a number of group-theorists have filled the gaps in the programme.

One striking result is the construction, by R.L.Griess, of the *Monster*, a simple group of order

$$2^{46}.3^{20}.5^9.7^6.11^2.13^3.17.19.23.29.31.41.47.59.71$$

whose existence was conjectured by Bernd Fischer.   A previously conjectured subgroup, *Baby Monster*, was constructed earlier by Sims and Leon.   The Monster is currently the subject of several wild (and probably justified) conjectures concerning possible links with complex analysis and modular functions, and it is clear that something pretty deep is going on - even if no one is sure what!

BIBLIOGRAPHY

J.H.Conway   *Monsters and Moonshine*, Mathematical Intelligencer 2 (1980) 165-171.

I.N.Stewart   *The Elements of Symmetry*, New Scientist 82 (1979) 34-36.

'Twas brillig, and the slithy toves
Did gyre and gimble in the wabe:
All mimsy were the borogoves,
And the mome raths outgrabe.

Lemmawocky

CAROL LEWIS

Thus, as educated persons know, runs the first verse of Lewis Carroll's *Jabberwocky*. He goes so far, in *Alice*, as to explain what the words mean. I find his explanation slightly evasive: it gives a meaning to each word separately, but not a coherent one to the entire stanza. Yet it clearly possesses a coherent structure, and hence a deeper significance - but not one readily explicable in *Alice*. Now Carroll was a mathematician - and what could be more difficult to explain to a Victorian child than mathematics?

Jabberwocky, then, is mathematical in content. The natural break at the end of line two of this first stanza suggests that it be interpreted as a

THEOREM Suppose (a) it is brillig, and (b) the slithy toves gyre and gimble in the wabe. Then all borogoves are mimsy, and the mome raths outgrabe.

It remains only to decipher this. It turns out that Carroll had anticipated a considerable part of what, until this article was published, has been thought of as very modern mathematics. This can only lead to a radical reassessment of Carroll's role as a mathematician.

Clue 1 is *wabe*. Carroll says it derives from 'way beyond' and 'way before'. The explicit avoidance of 'way below' makes it clear that *the wabe is the Euclidean plane* R^2. *Gyre* and *gimble* refer to motion: one is immediately put in mind of a dynamical system, a flow.

Slithy toves: what are they? We are told that *slithy* means 'slippery and slidy', so the toves must slide along the flow. They are manifestly the *tangent vectors* (see diagram), as borne out by the occurrence of the initial *t* and *ve* in both words. Indeed, one might speculate that the word was originally *tave*, short for *tangent vector*, but that the printer misspelt the word out of ignorance.

What was *brillig*? It can only be the dynamical system itself! So we can paraphrase Carroll's Theorem as follows. Consider a brillig dynamical system whose tangent vectors flow

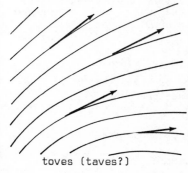

toves (taves?)

in the plane.   Then all of its borogoves are mimsy, and its mome
raths outgrabe.

I stood awhile in uffish thought, thumbing through Smale's sur-
very article [3] on differentiable dynamical systems... and found
a theorem of Peixoto [1]: any structurally stable system in R^2 has
finitely many singularities, and the α- and ω-limit sets of every
trajectory are singularities or closed orbits...

So *brillig* means 'structurally stable' and *borogoves* are singularities.   *Mome*, far-from-home, refers to limit sets of trajecto-
ries; *rath* is a misprint for 'path'.   *Outgrabe* is a verb which
must correspond to 'tend to a closed orbit or singularity', perhaps
grow out gradually (i.e. stay within a bounded set).   Now the whole picture emerges, and we render a tentative translation of the
entire poem.

'Twas brillig, and the slithy toves
Did gyre and gimble in the wabe:
All mimsy were the borogoves,
And the mome raths outgrabe.

Beware the Jabberwock, my son!
The jaws that bite, the claws that
    catch!
Beware the Jubjub bird, and shun
The frumious Bandersnatch!

He took his vorpal sword in hand:
Long time the manxome foe he
    sought -
So rested he by the Tumtum tree,
And stood awhile in thought.

And, as in uffish thought he stood,
The Jabberwock, with eyes of flame,
Came whiffling through the
    tulgey wood,
And burbled as it came!

One, two! One, two! And through
    and through
The vorpal blade went snicker-snack!
He left it dead, and with its head,
He went galumphing back.

And hast thou slain the Jabberwock?
Come to my arms, my beamish boy!
O frabjous day! Callooh! Callay!
He chortled in his joy.

'Twas brillig... (reprise)

A structurally stable dynamical system in the plane has
finitely many singularities,
and the limit sets are closed orbits or singularities.

The hero, a research stu-
dent, is cautioned by his
supervisor against pitfalls,
and the work of certain mathematicians, especially those
who might 'snatch' his ideas.
He sets to work, but meets
no success. He pauses to
let his subconscious function.

An idea forms and 'burbles'
up to the surface of his mind.
(Compare with Poincaré's [2]
discussion of the role of the
subconscious in mathematics.)

He realises how to solve
the problem by a method involving a double induction,
and cutting up the plane.
Having found a proof, he rushes off to tell his supervisor the main idea.

"You've proved it? Oh,
well done, lad! You'll get
a Ph.D. out of this!   And
I'll get my NSF grant renew-
ed!"
(Restatement of Theorem.)

It is intriguing to speculate on the advances that might have
been made, had Carroll's Theorem been deciphered earlier.

BIBLIOGRAPHY

1. M.Peixoto *Structural stability on 2-dimensional manifolds*,
    Topology 1.
2. H.Poincaré *Foundations of Science*, Science Press.
3. S.Smale *Differentiable dynamical systems*, Bull. AMS 73 (1967).

MANIFOLD *had a minor obsession with* Jabberwocky.
*A translation into Latin elegiacs, from the
Lewis Carroll Picture Book of 1899, was kindly
communicated to us by Gaberbocchus Press (we were
not alone), now associated with De Harmonie
(Singel 390, 1016 AJ Amsterdam - DO write for a
catalogue if you like the unusual). And so:*

Slythaeia Tova

the late Mr. HASSARD DODGSON   (after Lewis Carroll)

*Hora aderat briligi.   Nunc et Slythaeia Tova
   Plurima gyrabant gymbolitare vabo;
Et Borogovorum mimzebant undique formae,
   Momiferique omnes exgrabuere Rathi.*

'Cave, Gaberbocchum *moneo tibi, nate cavendum
   (Unguibus ille rapit.   Dentibus ille necat.)
Et fuge Jubbubbum, quo non infestior ales,
   Et Bandersnatcham, quae fremit usque, cave.'*

*Ille autem gladium vorpalem cepit, et hostem
   Manxonium longa sedulitate petit;
Tum sub tumtummi requiescens arboris umbra
   Stabat tranquillus, multa animo meditans.*

*Dum requiescebat meditans uffishia, monstrum
   Praesens ecce! oculis cui fera flamma micat,
Ipse* Gaberbocchus *dumeta per horrida sifflans
   Ibat, et horrendum burbuliabat iens!*

*Ter, quater, atque iterum cito vorpalissimus ensis
   Snicsnaccans penitus viscera dissecuit.
Exanimum corpus linquens caput abstulit heros
   Quocum galumphat multa, domumque redit.*

'Tune Gaberbocchum *potuisti, nate, necare?
   Bemiscens puer! ad brachia nostra veni.
Ah! frabuisce dies! iterumque caloque caláque
   Laetus eo' ut chortlet chortla superba senex.*

*Hora aderat briligi.   Nunc et Slythaeia Tova
   Plurima gyrabant gymbolitare vabo;
Et Borogovorum mimzebant undique formae,
   Momiferique omnes exgrabuere Rathi.*

Gaberbocchus.

*Topology.   To the modern mathematician, a powerful
and indispensable tool.   To many "practical" people,
a pointless abstraction.   But significant ideas are
never pointless; and lack of imagination is never truly
practical.   As the century unfolds, we are witnessing
the rise of:*

Topology in the Scientist's Toolkit

CHRISTOPHER ZEEMAN

Topology is a sort of geometry, so let me start right away with a
theorem.   Imagine a ball, covered all over with hair, and then try
to comb the hair down smoothly.   The theorem says that you can't -
just that.   There's got to be a tuft or a whorl somewhere.   For
example, a coconut is covered with hair and it has a tuft at the
top and a parting at the bottom.   Alternatively, if we tried to
comb the coconut round sideways, then there'd be a whorl at the top
and a whorl at the bottom.

I admit that this sounds a rather whimsical theorem, but it does
have practical applications, and it can be generalized to deep and
powerful theorems in higher dimensions.   But before we go into higher dimensions, let me mention a couple of applications.

The first one is about the weather.   There can never be a stable
weather situation with the wind blowing smoothly all over the Earth,
otherwise we could imagine the Earth to be our hairy ball and comb
the hair down in the direction the wind was blowing, which would be
a contradiction.   So there must always be some whorls or cyclones,
and consequently unstable weather.   The second application is in
nuclear energy.   If we are trying to create nuclear fusion, we
have got to contain the whole process in a magnetic bottle, and on
the surface of the bottle the magnetic field has to be everywhere
tangential to the surface.   Now if we had a spherical shaped bottle then we could comb the hair down in the direction of the magnetic field, which would again give us a contradiction.   So it's no
good having a spherical bottle.   Nor is it any good having an egg-
shaped bottle, because we can't comb a hairy egg smooth either.   We
must have a completely different type of bottle, because it must
have a surface that can be combed smooth.   For example, a doughnut
would do.   The surface of a doughnut, the topologists like to call
a torus.   In other words the torus is the same as a rubber ring or
inner tube.   Now a hairy torus can be combed smooth by merely running one's hand around it.   Therefore a magnetic bottle shaped
like a torus will be all right for nuclear
fusion.   Alternatively we could have a
knotted torus, got by tying the ends of a
piece of hosepipe in a loose knot and then
joining the ends together.   Another example of a bottle, which I don't think the
physicists have tried out yet, is obtained
by taking a ball and boring out a knotted

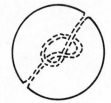

52

hole through it - the sort of hole that one would imagine a drunken woodworm might make by mistake.

So much for a magnetic bottle. But, as you can see, we've already begun to look at geometry from a completely fresh viewpoint. We are no longer interested in size, or straight lines, or in the difference between a sphere and an egg. We're interested in differences that seem much more basic, like the difference between a torus and a sphere. The torus has a hole and the sphere hasn't. The torus can be combed smooth and the sphere cannot. These differences are so basic that they persist even when the two surfaces are subjected to the most drastic type of deformation. And this is the secret of topology, and why it is sometimes called rubber-sheet geometry. Two objects are said to be topologically the same if, when we imagine them made of rubber, we can deform one continuously into the other. But no cutting or tearing is allowed. For example, we can deform a sphere into an egg, and so they are topologically the same. More surprisingly, we can deform a cube into a sphere: imagine the cube to be made of rubber and then pump it up. But a torus is not the same as a sphere because, however much we bend it about or pump it up, we can never get rid of the hole. Topology is the study of properties that are preserved under these continuous deformations. For example, whether or not a surface can be combed smooth.

Another good example of a topological property is the linking of circles. It's a favourite trick of conjurors to take a pair of metal circles and suddenly link them together. Of course they cheat by slipping one of the circles through a little hidden catch in the other. But the trick is always very effective because it shocks our intuition to the core. We know intuitively that even if two linking circles are made of string, then however much we waggle them about, they always remain linked. The intuition is very deep, because children acquire it before they can talk. I remember my son discovering he could link a red plastic ring on his arm one day in the bath when he was about six months old: both he and I sat entranced while he put it on and off his arm about twenty times. At that moment his intuition about linking was born. Sometimes our intuition creeps into our language: for instance we say that we've walked in a circle when we really mean that we've walked in a closed loop, which is topologically the same as a circle. And when we talk about inside and outside, we are intuitively affirming

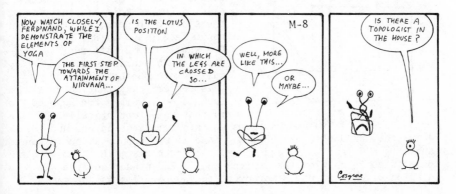

the famous Jordan Curve Theorem, which says that any closed curve
has an inside and an outside.

To the early mathematicians these 'facts' like curves having in-
sides and outsides or circles being able to be linked, were intuitively so obvious that they never thought of proving them.   And
when people came to prove them, the proofs turned out to be surprisingly hard.

Here's a nice example: construct three circles such that any two
are mutually unlinked, but the three together are linked.   It's
quite easy to construct the three circles...

... but how do we actually *prove* that they are linked?   Of course
the scientists would make them out of wire or string and fiddle about for a few minutes, and then say "there you are: they're linked",
and this would be a valid scientific proof - but no good as a mathematical proof, because it would be open to later refutation, if
some genius came along and showed us how to unlink them.   No, a mathematical proof must be a watertight argument holding for all
time.   And to show our three circles are linked, a topologist has
to use quite sophisticated group theory.

This, of course, is one of the delights of topology: the use of
sophisticated algebra to prove geometrical things.   It all began
with Euler in the 18th century, who discovered an interesting formula for polyhedra.   He showed that in any polyhedron like a pyr-
amid or a cube, the number of
corners, minus the number of edges, plus the number of faces,
always comes to 2.   Very remarkable.   For example, in the cube,
there are 8 corners, 12 edges,
and 6 faces; and so 8-12+6 = 2.
But what Euler didn't realise,
and what Poincaré discovered
just before the beginning of
this century, was that Euler's
formula only worked for polyhedra that were topologically equal
to a sphere.   If we chop a torus up into vertices, edges, and
faces, then we don't get 2 but 0.   And if we chop up a double doughnut, then we get -2.   And so on.   In each case, the number that
we get depends only on the surface and is independent of the way
in which we chop it up.   Moreover, the number remains the same

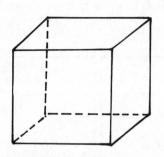

when we subject the surface to all these drastic topological deformations, and so we call it *invariant*.   In honour of Euler, Poincaré called this number the Euler invariant.

This is the secret of algebraic topology: the discovery and use of invariants.   If two figures have different invariants then we cannot deform one into the other, even if we try for a million years.   Consequently we can prove two figures topologically different. In this way we can prove mathematically that the sphere is different from the torus, and with another invariant we can prove that linked circles cannot be unlinked, and so on.   When Poincaré discovered the Euler invariant, he couldn't have dreamt of the wealth of invariants and theorems that have since sprung from that small seed. Poincaré's work around 1899 was so significant that topologists generally honour him as the father of the subject.

You may ask why we need this luxury of mathematical proofs for things whose scientific proof is so obvious.   In the three-dimensional world, it's a matter of taste which type of proof one prefers; but in four or five dimensions or up in 101 dimensions the mathematical proof is a necessity, because the scientific proof just doesn't exist.   It's in the higher dimensions that topology really gets interesting - because then we can prove theorems that we cannot quite visualize, and there are all sorts of surprises and new phenomena.   It's a mysterious and beautiful world.

But what are these higher dimensions, and why do we bother with them?   Let's agree right away that they are figments of our imagination - nobody is trying to say they exist in the same way that length, breadth, and height exist in our three-dimensional world. Nor are we trying to say that time is the fourth dimension, although in physics it is sometimes convenient to treat it so.   It's beast to look at it from another angle, the business of solving equations.

All sorts of scientific problems are continually coughing up equations to solve.   For example, suppose we are trying to make an economic model of the country: we might have a hundred and one variables: x_1 might be the gross national product, x_2 the price of petrol, x_3 the unemployment rate, and so on.   And all these variables and their rates of change might be connected by many equations and differential equations, and having got this huge long list of equations, what on earth are we to do with them?   We can't solve them because they are too complicated.   But we can attack them either quantitatively or qualitatively.   The quantitative approach is to put the whole thing on to a computer.   The computer will tell us the precise answer in any one instance or any number of instances. But it will not give us any intuitive grasp of the problem as a whole.   The qualitative approach on the other hand is quite different: it is geometrical and aims to give us a global picture, to see the problem in a Gestalt way, to visualize all the solutions at once and their interrelations with one another - without bothering about the detailed answer in any one instance.   We go about it as follows: for each variable x_1, x_2, etc. we invent a new dimension, and so we get a 101 dimensional space.   The equations then represent some geometrical figure in this space, or a *manifold* as the topologists like to call it.   The differential equations give rise to flow lines on the manifold.   If you like, you can think of the manifold as being hairy, and all the hairs lying down in the

direction of the flow lines. Any point on the manifold will represent a state of the whole economic system. And as time progresses, this point will flow along the flow lines. It may flow round
and round in a closed loop or it may spiral off to infinity, or it
may flow to an attractor point where it settles. Going back to
the familiar homely examples of manifolds in three dimensions, namely the surfaces that we had before, the flow on a torus would go
round and round in closed loops with no attractor points. The
flow on a coconut would be all towards the tuft at the top which
would be the only attractor point.

So you can see that these whimsical theorems about hairy surfaces
do give our intuition something to cling on to while we grapple
with the flow on a 101-dimensional manifold. In our economic
problem the flow round a closed loop might represent periodic booms
and recessions, and the spiral off to infinity might represent chro-
nic inflation, and the attractor point might represent a steady but
stagnant economy. Qualitatively we would want to understand the
manifold as a whole, so as to avoid these perils and steer the economy into a safe flow pattern.

So much for the economy. Let's consider another example: the
famous three body problem, the study of the motions of the Sun,
Earth, and Moon. It was probably this problem that originally sti-
mulated Poincaré to invent topology in the first place. Nowadays
it has become a pressing practical problem of how to put a satellite
into orbit between the Earth and Moon. Mathematicians have become
a dab hand at the quantitative approach, using computers to predict
particular orbits - but as yet they still don't properly understand

the qualitative aspect.   Each of the three bodies has three coordinates of position and three of momentum, so that altogether we
have 18 variables, and so it's a geometrical problem in 18-dimensio-
nal space.   If we take coordinates at the centre of gravity of the
three bodies, this has the effect of removing 6 dimensions and
brings us down to 12; if we fix the energy level this brings us
down to an 11-dimensional manifold.   Each point of the manifold
represents a state of the dynamical system, and as time progresses
the point flows along flow lines in the manifold, and represents
the way the three bodies revolve around one another.

I've described these two examples in economics and astronomy to
illustrate how often problems in many dimensions crop up in mathematics.   The topologist, however, is generally less interested in
the applications than in the manifolds themselves.   He likes to
study the manifolds, to classify them, and investigate the number
of different dimensional holes they have.   He likes to classify
the way they can be linked or knotted, and embedded in one another.
He studies flows on manifolds and classifies the different types of
flow lines.   Since 1960 we have witnessed the greatest development
in the theory of manifolds that there has ever been.

In a way it's art for art's sake: it is the intrinsic elegance
of the subject rather than the applications that dictates the way
the subject grows, and the direction of research in topology.
Like all branches of pure mathematics, it paradoxically becomes
both simpler and more complicated at the same time.   Every day we
discover more and more theorems, and so topology becomes more complicated; but at the same time new underlying patterns reveal themselves and so it becomes simpler.   We develop a confidence in han-
dling problems that were previously beyond Man's comprehension.   At
first topology was strictly for topologists, but now it is rapidly
becoming part of the toolkit of many scientists.

Personally, one of the most exciting developments that I forsee
in the next twenty years will be the increasing use of topology in
biology, and even the creation of a subject called theoretical biology analogous to theoretical physics.   Up to now biology has needed to use very little mathematics, because most of the interesting
experiments could be done without bothering with it (apart from a
little statistics).   But now biology is hitting problems that may
need really sophisticated mathematics before we understand them.
Let me mention two examples in particular.   The first is the global
activity of the brain.   I have worked on this myself, in terms of
the flow on a 10,000,000,000-dimensional manifold.   The second
area is in morphogenesis - how the egg develops into an embryo and
eventually into an adult animal.   The structural stability of species, that is to say why all the animals in one species more or
less look alike and have the same organs, is the greatest unsolved
problem in biology today.   It's all very well to say that we have
discovered the code of life, the sequence of DNA molecules in the
genes; but nobody yet has a clue to how this code actually works.
Eventually we have got to explain how the code instructs the embryo
to grow and develop folds and to differentiate into the multifari-
ous different organs of the body.   One of the leading pioneers in
describing these changes (not explaining them yet) has been the em-
bryologist C.H.Waddington of Edinburgh.   Today we witness a fasci-
nating development: one of the world's greatest topologists, René

Thom of Paris, has taken up Waddington's work and applied topology to biology to push the description one stage further towards an explanation. He has written a book shortly to be published.

Thom describes the chemistry of the embryo in terms of a high dimensional manifold, and visualizes the chemical changes being controlled by a flow pattern which is itself caused by chemical gradients set up by the genes. At each point of the embryo the chemical state is flowing towards its attractor. So, as Thom rather poetically puts it, the whole embryo is being blown towards its attractor surface by the metabolic wind. At the same time, and the understanding of this point is one of Thom's strokes of genius, the flow causes the attractor surface itself to develop folds or singularities. It is these folds which cause the precise nature of the morphogenetic changes such as limb bud growing into two bones. The surprising thing is that there are only seven possible types of fold. So these seven types are responsible for all simple growth changes in all living things. Suddenly amidst the apparent complexity of life, we glimpse just a few simple controls. It is a classic example of a piece of knowledge developed by pure mathematicians for its own sake, and suddenly used to explain something that had previously looked quite hopeless. It pierces to the heart of the matter.

* * * * * * * * *

HINT 2

DO NOT READ THIS NOTE UNTIL YOU HAVE READ PAGE 45.

"There are of course Watson, two possible mating moves, both requiring some knowledge of how the play has proceeded. There are a number of potentially useful observations that we can make immediately."
"I fail to see anything useful, Holmes."
"I commend to your attention the strange affair of the Bishop on f1, Watson."
"But there is no Bishop on that square!"
"Precisely, Watson! We may then ask rather pertinently... where did the Bishop on d5 come from, then?"
"Holmes! And the Bishop on e5 also?"
"Indeed, Watson."

The Bishops are of course promotees. Bishops do not change the colour of their squares during play, and neither of the Bishops on squares f1 and f8 at the start of the game can have moved. Also, the Bishop now on e5 can easily be seen to have been the promotion of a pawn that captured as it was promoted...

The next hint is on page 84.

* * * * * * * * *

SIGNS A MANIFOLD reporter, walking through one of our more established universities, came across a sign bearing the legend: CONTROL ENGINEERING. Being the purest of pure mathematicians, he heartily agrees!

M-3

It all began with Von Neumann and Morgenstern's
'Theory of Games and Economic Behaviour'.   But
in mathematics, as in real life, many questions
of strategy more resemble:

A Pandora's Box of non-Games

ANATOLE BECK and DAVID FOWLER

In the study of Games, as in many other intellectual pursuits, one
of the important problems is to Find The Question.   When the question has been found, the answer may be sought in good time.   In
Game Theory, there are simple games, like the matrix games, and
very, very complex games.   Today, the centre of Game Theory is occupied by the theory of cooperative games, in which it is not yet
known what an answer would be, much less how to find one.   We will
include below a few simple cooperative games.   In addition, we
will exhibit some things which are almost certainly games, except
that they are so ephemeral, so indistinct, that they still defy analysis.   Unlike Chess or Go, where the complexity arises from the
multiplicity of possible strategies, the difficulty here arises because of the great simplicity.   No doubt if the games were more
complex, the difficulties would be hidden.   Let's start with an
English game:

1. *Finchley Central*.   Two players alternate naming the stations on
the London Underground.   The first to say 'Finchley Central' wins.
It is clear that the 'best' time to say 'Finchley Central' is exactly before your opponent does.   Failing that, it is good that he
should be considering it.   You could, of course, say 'Finchley Central' on your second turn.   In that case, your opponent puffs on
his cigarette and says 'Well,...'.   Shame on you.

2. *Penny Pot*.   Players alternate turns.   At each turn, a player
either adds a penny to the pot or takes the pot.   Winning player
makes first move in next game.   Like F.C., this game defies analysis.   There is, of course, the stable situation in which each player takes the pot whenever it is not empty.   This is a solution?

Penny Pot has an interesting variant:
3. *Penny Pot with Interest*.   The Pot is a bank account, on which
the players draw interest, which they share.

The next game is a three-person symmetric game.
4. *Lucky Pierre*.   Each of the three players chooses a positive integer.   If all three numbers are different, then the one in the
middle collects a franc from each of the others.   If two are the
same, the odd man out collects.   If all are the same, then no dice.
This game has some interesting analysis.   If two of the players
gang up on the third, then they can take 4 and 5.   No matter what
happens, one of them wins.   They share the loot.   To make the

game have meaning, there has to be some sort of real bar to collusion. If there is, and if all the players are thought of as intelligent (where did that hypothesis come from?), then we have the following chain of theorems.

Theorem 4.1 No one ever plays 1.
Proof   There is almost no hope of winning when you play 1. Only if the other two tie, can you get *anything*. And then not much. QED (?).

Theorem 4.2 No one ever plays 2.
Proof   Since no one ever plays 1, by the previous theorem, the same reasoning applies to 2. QED (??).

Theorem 4.3 No one ever plays 3.
Proof   Obvious (?????????)!

There are other theorems, too numerous to mention, and which together imply:

Theorem   No one ever plays.
Proof   Left to the reader as an exercise (!!).

This analysis is similar to that of the Surprise Examination, for those of you who know it.   Another game with the same analysis is:

5. *Big Number*.   Two players.   Each chooses a positive integer. The owner of the smaller integer pays a rupee to the owner of the larger.   Theorems 5.1, 5.2, 5.3, etc. are left for formulation to the reader.

Big Number differs sharply from
6. *More Money*.   Two players with literally infinite resources. Each bets an amount of money.   The larger bettor wins the stake of the smaller (see Marx, *passim*).   Here, it is common to see players betting pennies, except from time to time.   Similar to F.C. and Penny Pot, in some ways.   The object, of course, is to win, rather than come out winning.   Only when you have infinite resources can there be a distinction between these.

7. *Come to Dinner*.   Two players, Source and Sink.   Mr. Source offers dinner to Mr. Sink ('Come to dinner').   Mr. Sink refuses, indicating that he would like dinner, but courtesy forbids (e.g. 'It is late, and your wife is not expecting me').   Source insists ('We have Stroganoff tonight, and Denise always makes plenty').   Sink ducks again.   Finally Source says 'Very well, some other time.' Or Sink says 'All right, since you insist'.   Whoever says this line WINS.   The game is played for two prizes, Dinner and Honour. The principal object is to get (resp. avoid giving) Dinner, and to do so while obtaining as much Honour (measured in rounds) as possible.   Both players accrue Honour, but no amount of Honour can compensate for the loss of Dinner.   The payoff is non-Archimedean. Note the similarity of this game to F.C. and P.P.

8. *Tweedledum and Tweedledee*.   The Red Queen offers 1000 marks to Tweedledum and Tweedledee if they will agree how to share it; time limit.   Tw-m offers 50-50, but Tw-e holds out for DM 650:

Tw-m: If you don't take this, you'll get nothing.   Come on, it's DM 500 or nothing!

Tw-e: Same goes the other way: do you want the DM 350 or not?

If you think this is a simple game, imagine the inventor who has a device which can save the telephone company $ 1,000,000 a year.

The device is patented, and no one but the telephone company can use it. How much is his share of the take?

9. *Winken, Blinken, and Nod.* Same Red Queen, having failed to dispose of her swag to Tw&Tw Ltd., offers same to Wn, Bn + N on the following terms. If they all will agree on the mode of sharing, they get DM 1000. If two agree without the third, they get the following, depending on which two they are; odd man gets nothing: Wn+Bn 500, Wn+N 750, Bn+N 600. No agreements mean no dough. Who gets how much? Shame on you if you don't hit the patsy for the whole grand.

Tw&Tw and Wn,Bn+N are called cooperative games. If you think these are screwy, you should see what happened the day the Red Queen offered some cash to Haupt, Voll, Blut, and Wunden.

10. *An Infinite Game.* A real mathematician's game with a real mathematician's solution. A and B alternate choosing positive ($\neq 0$) real numbers to form a decreasing sequence; they play forever. At the Trump Of Doom they add up their choices (infinitely many). If the sum is infinite or rational, A wins. Otherwise B. How does it come out?

$$* \quad * \quad * \quad * \quad * \quad * \quad * \quad * \quad *$$

The ubiquitous Eve LaChyl gave the answer in M-4: *B wins. This is because the rationals are countable. By making his choices decrease faster than* 2^{-n}, *B can ensure convergence. By imposing other conditions he rules out the n-th rational in an enumeration as a possible sum. Details left to the reader.*

The article provoked some correspondence, duly recorded in M-4. *Additional games included:*

Miss Take. A panel of judges view a selection of distinct but undistinguished young women. This preliminary is traditional but unnecessary to what follows, when the judges each nominate one of the women as being the one most likely to be chosen by all the other judges. All judges have this criterion, and no other, in mind.

One Arm Band-Aid. A solitaire game. The player pays sixpence to pull a handle, when pretty coloured symbols flash before his eyes. There is no other penalty for losing. No one ever wins.

In M-5 *M.Henton of New Addington noted with horror that there is an isomorphism between Finchley Central and the game commonly known as 'Nuclear Deterrent'. "It occurs to me that we should work very fast to analyse the non-games, before we are left with a non-world."*

Which reminds us of Larry Markus's favourite non-game. Two players play chess (normal rules). Afterwards, they toss a coin to see whether checkmate means 'win' or 'lose'. There is again a real-world isomorph, called 'Advanced Technology'. There is another one, called 'Lack of Advanced Technology'.

$$* \quad * \quad * \quad * \quad * \quad * \quad * \quad * \quad *$$

The longest known sequence of primes in arithmetical progression, consisting of 16 primes, has been found by S.C.Root of Massachusetts. The first in the progression is 2236133941, and the common difference is 223092870. (As a check, the 16th prime is 5582526991.)   M-6

*A proof, in practice, is seldom a series of logically
connected steps like the textbooks say.   Life is too
short.   The main point is to be convincing.   There
are many ways to add conviction to an argument: we
look at one that is widely popular...*

the MANIFOLD guide to Handwaving

Cosgrove

THE ONLY
METHOD TO
PROVE THIS
THEOREM...

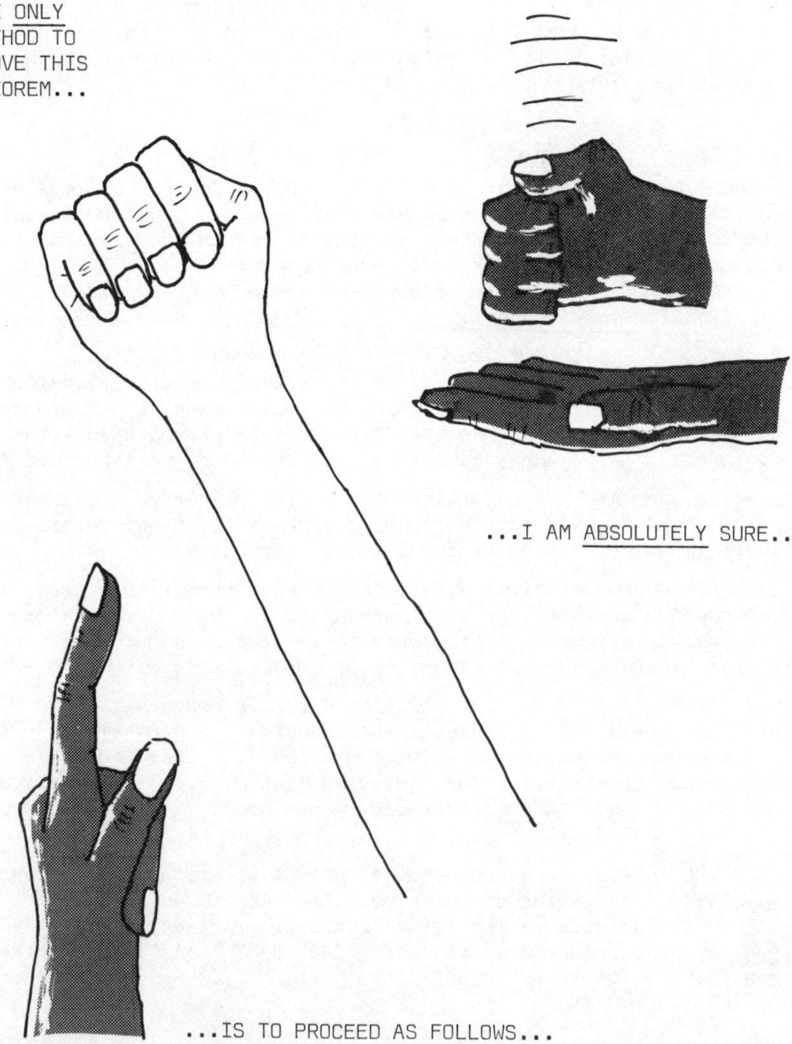

...I AM ABSOLUTELY SURE...

...IS TO PROCEED AS FOLLOWS...

BUT BEFORE GOING
ANY FURTHER...

...WE MUST CHECK THE
HYPOTHESES OF THE
LEMMA ARE
SATISFIED!

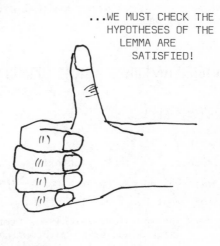

UNFORTUNATELY,
THEY AREN'T.

...BUT WE CAN EASILY
AVOID THIS DIFFICULTY...

...BY MAKING
SOME...

...MINOR
ADJUSTMENTS
...

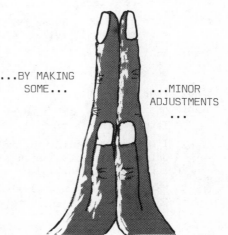

63

Despite - or possibly because of - their excursion into Maze Theory, Theseus and the Minotaur have not managed to travel more than the odd mile or so from the centre of the Labyrinth; and they now tend to spend their time in mathematical pastimes. We eavesdrop once more...

Meanwhile, back in the Labyrinth...

STEVEN EVERETT

Theseus: 1152!

Minotaur: Was that a factorial or an exclamation mark?

Thes: Oh, an exclamation mark - I'm nowhere as high as 1152!

Mino: What are you doing, then?

Thes: The four fours problem - trying to represent numbers by four 4s combined using mathematical symbols. 1152 is 4!4!+4!4! - and I do mean !

Mino: (Checking first) Yes, that's right. That doesn't seem bad at all - isn't 1152 terribly good?

Thes: Well, it would be if I had all the numbers up to 1152, but so far I've only got an unbroken sequence as far as 873.

Mino: What's the highest that you have?

Thes: 1152.

Mino: Surely you can keep going to as high a number as you please? I mean, isn't 1152! just (4!4!+4!4!)! and so on - just adding factorial signs?

Thes: (Slowly - which gives you an idea of the abilities of this pair) Yes.

Mino: 873 seems reasonable, though.

Thes: It's not bad. Rouse Ball won't be able to do any better than this until 1912.

Mino: Who's Rouse Ball?

Thes: Oh, he's someone who hasn't been Bohrn (sic) yet!

Mino: What's *sic* mean?

Thes: Oh, that's to mark a horrible pun, because of what's coming later in this script.

Mino: I'm doing all right so far - I'm as far as 12, but I can only use three 4s - either as 4+4+4, or as 4×4-4. How can I get to use four?

Thes: Well, you could use $\sqrt{4}\times\sqrt{4}\times4 - 4$, but there's another way which is just to multiply one of your answers by 1, using only a single 4. One way of getting unity with a single 4 is $1 = [\sqrt{\sqrt{4}}]$ where the brackets mean *integer part of*.

Mino: Ah, I see - that's quite good.

(There is a pause - Minotaur is clearly working through the integers.)

Can you do 1153?

Thes: No, not yet.

(Pause.)

Mino: I can: look! $1153 = -\sqrt{4} \log_4 \log_4 \sqrt{\sqrt{\sqrt{\sqrt{\sqrt{\sqrt{\sqrt{\sqrt{}}}}}}}}\ldots\sqrt{\sqrt{\sqrt{4}}}$.

Thes: I'm sorry, I wasn't following that too closely - could you be more precise?

Mino: Mmmmm - there were 1153 square root signs, in case you didn't count.

Thes: It's still a bit unclear, though.

Mino: Well: $\sqrt{\sqrt{\sqrt{\sqrt{\sqrt{\sqrt{\sqrt{}}}}}}}\ldots\sqrt{\sqrt{\sqrt{4}}}$ is 4 to the power $1/2^{1153}$. OK?

Thes: Yes - so that $\log_4$ of it is just $1/2^{1153}$!

Mino: That's right, and $\log_4$ of *that* is...

Thes: ... $-1153/2$ which gives, on multiplying by $-\sqrt{4}$,...

Mino: 1153!

Thes: You know something - I can do 1154.

Mino: And 1155, and 1156,...

Thes: We've pretty much sewn up the integers, haven't we?

Mino: We have, rather - have we reached the sic joke yet?

Thes: Oh, yes - this is going to be invented by Niels Bohr in the 20th century after Christ.

Mino: Who's Christ going to be?

Thes: Ah, shaddup! What are we going to do now? There's nothing left of the integers.

Mino: Well, we could try some irrational numbers - π or e, for example.

Thes: Oh, that's transcendental, man! Anyway, I can do e, sort of.

Mino: What do you mean, sort of?

Thes: Well, $(1+\dfrac{1}{4.4!!!!!!\ldots})^{4.4!!!!!!\ldots}$ is a good approximation to e, but you have to take an infinite number of ! signs.

Mino: I see - you're looking at the limit as n tends to ∞ of $(1+\dfrac{1}{n})^n$. Gee, Theseus, you are clever. NOW what will we do?

Thes: π probably - but let's leave that to the reader!

BBC Television has a very popular programme called
The Multicoloured Swap Shop. Below, MANIFOLD pre-
sents the Multicoloured Theorem Shop, running the
gamut of the integers from 1 to 5...

the 1-colour Theorem

JOZEF PLOJHAR

The little-known one-colour theorem is due to the persistence of a
long-since forgotten cartographer of about 3 years of age, who
like all children of such an age covered his maps in a single wash
of colour.   His father immediately realised the significance of
this, and burst into the mathematical journals with:

THEOREM   *All coloured maps are coloured with a single colour.*

Before reprinting the proof, a comment is surely due on the power
of this theorem - unlike later chromatic theorems, it does not ass-
ert that the map *may* be so coloured, but rather, that it IS!

Proof   (By Mathematical Induction).   Let $P(n)$ be the proposition
"all maps with n regions are 1-coloured".
   $P(1)$ is trivially true.   We show that $P(n)$ implies $P(n+1)$.   Consider a map with n+1 regions.   Remove one region: by induction the
resulting map is 1-coloured, with colour C, say.
   Again consider the map with n+1 regions, but now remove a different single region.   The remaining n are 1-coloured, with colour K,
say.   But K = C as there are some regions which have been 1-colou-
red twice (if you see what we mean!).
   Hence all n regions are coloured with colour C, and $P(n+1)$ is
true.   Hence, by induction, $P(n)$ is true for all n - and all maps
are one-coloured!

the 2-colour Theorem

VIVIENNE HATHAWAY

The infamous 4-colour problem asks, as you are no doubt aware, whether any map on the plane can be coloured using 4 colours so that
no two adjacent regions have the same colour.   It is not our pur-
pose to go into this question here: we set our sights lower and aim
at a theorem about *two* colours.   Until MANIFOLD produces a colour
supplement, this is more appropriate!
   The theorem occurred to me some years ago, but subsequent delv-
ing into the literature revealed that it is well known.   The result is this:

Suppose a finite number of circles is drawn on the plane. Then the resulting map can be coloured with two colours so that adjacent regions have distinct colours.

The theorem generalizes to closed curves rather than circles:

How do we go about *proving* such a theorem?   If we *try* colouring a given circle-map, it becomes clear that as soon as one region is coloured the rest follow automatically.   A proof based on this would have to show that all regions are reached, and that there are no contradictory choices of colour.   This boils down to considering circuits of regions (regions each touching the next along an edge, starting and finishing at a given region).   Only if all such circuits contain an even number of regions will the method work. They do, and it does, but that's not the best way to prove the theorem!

a circuit of 8 regions. Is the number always even?

It occurred to me that the theorem is accessible by Mathematical Induction on the number n of circles.   If n = 1 it is easy to colour the map:

So now we assume we can two-colour any map with n circles, and try to prove that we can two-colour a map with n+1.   Now any n+1 circle map comes by adding a circle to an n-circle map.   The diagram on the next page is typical.   If we can work out how to perform ? in general, we can prove the theorem.   Inspection of the diagram reveals that:
(a) Each region outside the new circle retains its colour,
(b) Each region inside the new circle changes colour.
   To see that this works in general, note that the colours

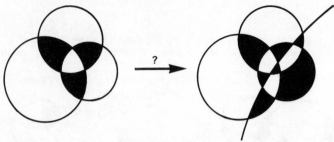

obviously change across boundary-lines outside or inside the new
circle (as they did on the old map: though inside, the colours are
reversed). And across the new circle's boundary, what was once a
single-coloured region divides into two, of opposite colours. QED!
   This is all very well - though not a very practical way to perform the colouring - try it - but by analysing the proof, we can
find something better.   Every time we add a new circle, *points
inside it change colour*.   So points inside an odd number of circles end up black; points inside an even number (or none) white.
This gives us a rule: *assign to each region an integer, equal to
the number of circles that contain it.   If this number is even,
colour the region white; if odd, colour it black*.   It is obvious
that this number changes by 1 from a region to the next: this gives
an independent proof.   Here's an example of the rule in operat-
ion:

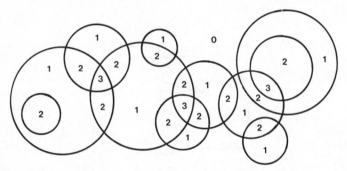

The proof obviously generalizes - e.g. to convex curves rather than
circles.   If a moral to the tale is needed, it is presumably that
our first ideas of how to solve a problem, based on direct solutions of special cases, may not be the best way to proceed in general;
and that analysis of a successful method can lead to improvements.

the $[(7+\sqrt{1+48p})/2]$-colour Theorem

Sorry about that.   That's what the *Heawood Conjecture* suggests as
the *precise* bound on the number of colours needed for maps on a
surface of genus $p \geq 1$ (a torus with p holes, or its non-orientable
analogue).   It was proved by Ringel and Youngs in 1968.

the 3-colour Theorem

Now, that's a puzzle!   MANIFOLD-12 set it as a competition: *Find
the 3-colour theorem*.   Although there *are* standard 3-colour theorems in graph theory, they all have rather artificial hypotheses.
We're still waiting...

the 5-colour Theorem

was until very recently the best that was known towards that doyen
of mathematical intractability, the 4-colour *problem*.   Which could
have caused us headaches with headlines (in contrast to our customary neckaches with necklines...).   Fortunately, two Illinois mathematicians, assisted by a whacking great computer, arrived in the
nick of time with:

the 4-colour Theorem

DOUGLAS WOODALL

In July of 1976, K.Appel and W.Haken, two mathematicians at the
University of Illinois in America, announced the solution of what
was probably the best-known unsolved problem in the whole of mathematics: the four-colour map problem.   This asks whether the regions of a map can always be coloured with four colours in such a
way that no two neighbouring regions have the same colour.   (Neighbouring here means 'having a length of common border'.   We do not
insist on giving two regions different colours if they meet only at
a finite number of points, like regions D and F in Fig.1.)

Fig.1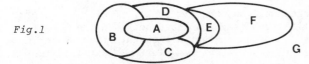

This problem was first proposed in 1852 by a London student, Fran-
cis Guthrie, who is reported to have thought of it while colouring
a map of the counties of England.   He noticed that four colours
are sometimes needed (e.g. for regions A,B,C, and D in Fig.1) and
conjectured that four colours always suffice, but was unable to
prove this.   The first serious attempt at a proof seems to have
been made in 1879 by A.B.Kempe, a barrister and keen amateur mathematician who later became President of the London Mathematical Society.   In that year he published a 'proof' in the American Journal of Mathematics which seems to have been generally accepted.
But in 1890 P.J.Heawood, Professor of Mathematics at Durham,

pointed out that the 'proof' contained a flaw. For some years after that the flaw seems not to have been regarded as serious, and
the theorem was thought to be 'essentially proved'. However, as
the years went by and nobody found a satisfactory way round the difficulty, it gradually became realised that the problem was much
deeper than had been supposed. Since then, almost every mathematician of repute has probably dabbled with the problem at some time
or other, so Appel and Haken's achievement in solving it (in the
affirmative) is a very fine one.

As might be expected of such a refractory problem, the proof is
long. It runs to 100 pages of summary, 100 pages of detail, and a
further 700 pages of back-up work, plus about 1500 hours of computer time. (For comparison, the average proof presented in first
year lectures probably does not last more than one page. In the
published literature I would regard a 20-page proof as quite long.)

Preparatory Moves

In common with most recent workers, Appel and Haken tackled the
problem in the form 'show that the vertices of every planar graph
can be coloured with four colours so that no two adjacent vertices
have the same colour'. A *planar graph* is a graph (= network)
drawn in the plane without edges crossing: see Fig.2. It is easy
to show that this version is equivalent to the original map problem
(stick a vertex in the middle of each region of the map, and join
vertices whose corresponding regions are adjacent). It is also
easy to show that it suffices to consider *plane triangulations*, i.e.
graphs that divide the plane into regions bordered by exactly three
edges (can you see why?). Fig.2 shows the graph corresponding to
the map in Fig.1, and the same graph made into a triangulation.

Fig.2.

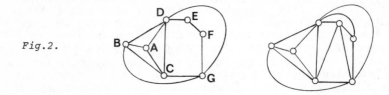

Kempe's "Proof"

In order to understand Appel and Haken's proof, it will be helpful
to start by translating Kempe's attempted proof into the language
of plane triangulations. Kempe started with Euler's polyhedron
formula, which states that a plane triangulation T satifies the
relation $V-E+F = 2$, where V,E,F are the number of vertices, edges,
and faces (regions) of T. (Can you prove this?) Since every
face of a triangulation is bordered by three edges, and every edge
borders two faces (the "outside" is thought of as one huge face),
we must have $2E = 3F$ (why?). If V_i denotes the number of vertices
of valency i (the *valency* of a vertex is the number of edges incident with it) then clearly $\Sigma V_i = V$ and $\Sigma i V_i = 2E$ (every edge has 2
ends). Substituting these in Euler's formula now gives

$$\sum (6-i)V_i = 12 \tag{1}$$

or, more longwindedly,

$$4V_2 + 3V_3 + 2V_4 + V_5 - V_7 - 2V_8 - 3V_9 - \ldots = 12.$$

It follows immediately that at least one of V_2, V_3, V_4, and V_5 is positive; so T must contain at least one of the four configurations in Fig. 3.

Fig.3   (a)   (b)   (c)   (d)

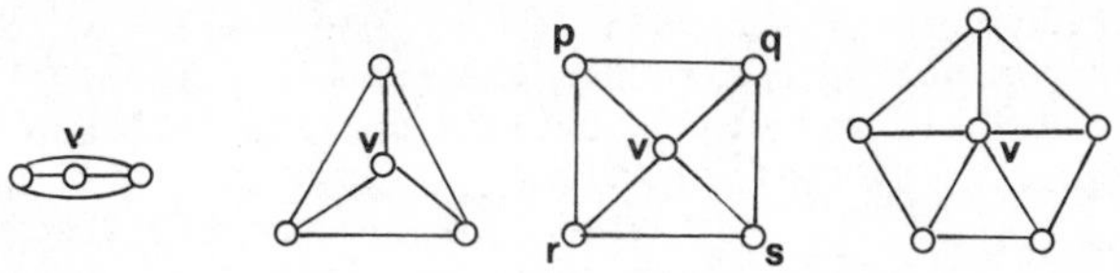

Now suppose there exists a counterexample to the 4-colour conjecture, and let T be a triangulation that is a minimal counterexample, so that every graph with fewer vertices than T *is* 4-colourable, but T itself is *not*. We naturally hope to prove this is impossible by obtaining a contradiction.

If T contains Fig.3(a) or 3(b), we need only remove v from T (together with the incident edges), 4-colour what is left, and restore v: since v is adjacent to at most 3 vertices, we can find a colour for it. Thus we have 4-coloured T, a contradiction. So T cannot, in fact, contain 3(a) or 3(b).

For 3(c) we try the same thing, but this time we are in trouble if p,q,r, and s all have different colours; in this case we cannot colour v. However, Kempe ingeniously showed, using what is now called a *Kempe-chain argument*, that here we can modify the colouring scheme so that either p and r, or q and s, have the same colour. Then we can find a colour for v, and again obtain a contradiction. (You can probably see how this can be done. If p,q,r,s are blue, green, red, and yellow respectively, then the graph T with v removed cannot contain *both* a chain of connected vertices from p to r, all blue or red, *and* a chain from q to s, all green or yellow; for these chains have to cross somewhere, and they can't.) Thus Kempe showed that T cannot contain 3(c) either.

If he could have shown that 3(d) was also ruled out, he would have completed his proof. Unfortunately, he tried to use the same trick for 3(d) as he had for 3(c), and thereby made his mistake, because the argument breaks down.

Nevertheless he made a very fine contribution towards the solution of the problem, often underestimated by later writers. Although his "proof" was fallacious, and hence technically worthless, the slightest modification of his argument yields a valid demonstration that *five* colours suffice; and his arguments have formed the foundation for most subsequent work on the problem.

The two main steps

To summarize Kempe's argument in modern terminology, he attempted to exhibit a set U of configurations (3(a)-(d)) such that:

(i) U is *unavoidable*: every plane triangulation contains one of the configurations in U;

(ii) Every configuration in U is *reducible*: it cannot be contained in a minimum counterexample to the 4-colour conjecture (i.e. any counterexample containing it also implies the existence of a smaller counterexample).

If his attempt had succeeded, it would certainly have provided a proof.   It failed, because he did not show satisfactorily that (d) is reducible.   Appel and Haken have been successful with exactly the same approach.   But while Kempe's unavoidable set contained 4 configurations, theirs contains about 1930.   (I say 'about' because they keep managing the reduce the number by 1 or 2.)   The proof that these are all reducible involves massive reliance on the computer.   One of their configurations is shown in Fig.4, and it is

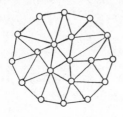

Fig.4

bordered by a circuit of 12 edges.   All of their configurations are bordered by circuits of 14 or fewer edges.   If they had used configurations larger than this, they would probably not have been able to prove them reducible with the present generation of computers.

Appel and Haken's proof thus involves the above two steps: the construction of U, and the proof that everything in U is reducible.   Each step is comparatively straightforward on its own: it is the interplay between them that is sophisticated, and in which Appel and Haken's work goes qualitatively, and not just quantitatively, way beyond anything that had been done before.

Contruction of an Unavoidable Set

To illustrate the first step, we show how Appel and Haken's method proves the set of configurations in Fig.5 unavoidable.   The idea is due to Heesch.

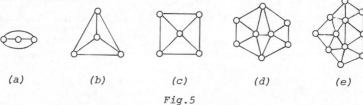

(a)   (b)   (c)   (d)   (e)

Fig.5

Suppose there exists a triangulation T not containing any of these. Assign to each vertex of T of valency i the number (6-i).   Appel and Haken like to think of this as (6-i) units of electrical charge; so a 5-valent vertex receives charge +1, a 7-valent vertex charge -1, an 8-valent vertex charge -2, and so on.   By (1), the total charge is positive (12 units).

We now redistribute the charge round T, without creating or destroying any, according to the following simple *discharging algorithm:* move 1/3 unit of charge for each vertex of valency 5 to each adjacent vertex of valency 7 or more.   T still has positive total charge.   But it is easy to check, using the fact that T contains none of 5(a)-(e), that no vertex of T can have positive charge!   For T has no vertex of valency $\leq$ 4; any vertex of valency 5 is adjacent to at least three of valency 7 or more, so loses all its unit of positive charge; vertices of valency 6 are unaffected, ending up with charge 0, where they began; a vertex of valency 7 can have at most three neighbours of valency 5 (or two of them would be adjacent) and so recieves at most 1 unit of charge, remaining negative; and so on.   This is a contradiction, so T must contain one of 5(a)-(e).

72

(Strictly speaking, this does not prove that one of 5(a)-(e) occurs
in T *with all of its vertices distinct*.   It is easy to get round
this for small configurations, but for larger ones it is a serious
technical problem, the *immersion problem*, and Appel and Haken had
to deal with it.)
    Appel and Haken proved their much larger set U unavoidable in
this way, but using a more complicated discharging algorithm.

Reducibility

To illustrate this step we again take an example, showing that 6(a)
is reducible.

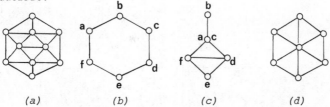

(a)   (b)   (c)   (d)

Fig.6

Let T be a triangulation that is a minimal counterexample to the
conjecture, and suppose T contains 6(a).   Let T' be the graph obtained from T by removing the four vertices inside the hexagon in
6(a); that is, replace 6(a) by 6(b).   By minimality of T, T' is
4-colourable.   List the possible colour schemes for the vertices
abcdef.   There are 31 of them:

```
121212   121213G   121232   121234G   121312G   121314   121323G
121324G  121342G   121343G  123123    123124    123132G  121313
123134   123142    123143   123212G   123213G   123214G  123232G
123234   123242    123243   123412    123413    123414G  123423
123424G  123432G   123434G.
```

(Here the numbers 1234 are the colours, listed in order on vertices
abcdef.   The G will be explained below.   Note that 121211 and
121231 are not listed, since they give adjacent vertices the same
colour (1); and 121214 is not listed since it comes from 121213 by
permuting colours.   Possibly not all of these can actually occur
in T', but we don't know which do, so we have to consider them all.)
    Some of these colour schemes can be extended to colourings of
6(a), so giving rise to 4-colourings of T.   Call these *good* (which
is what the G stands for).   If all colour schemes are good, then
6(a) is clearly reducible (because we can 4-colour T, a contradiction).   However, this never happens in practice.
    The next step is to try to use Kempe-chain arguments to convert
bad schemes into good ones.   For example, 121232, which is bad,
can always be converted by [13][24] interchanges into one of 121434,
121234, 121432, or 123232 - all good.   If every bad colour scheme
converts to a good one like this, then again 6(a) is reducible: we
say a configuration that can be proved reducible this way is
D-reducible.
    The first thing the computer checks for is D-reducibility.   (You
should now see why the size of the outer ring is crucial!)   If not,
the next step is to note that we don't actually have to consider
all 31 schemes on the list.   By minimality, we can replace 6(a) by

any configuration with fewer vertices, such as 6(c): the result T''
must be 4-colourable.   The effect of this substitution, here, is
that we need consider only colour schemes where a and c have the
same colour, and d and f are different: this rules out all but 6
of the schemes listed. If (as in this case) all the remaining schemes are good, or can be made so by Kempe-chain modifications,
then again we get reducibility.   There are many choices in place
of 6(c) - another example is 6(d), which shows we need consider
only schemes using 3 or fewer colours on abcdef.   If any such
substitution works, we call the original configuration *C-reducible*.

   The program used by Appel and Haken, largely written by a post-
graduate student John Koch and using algorithms of H.Heesch, first
checked for D-reducibility; if this failed, it tried a few ways of
proving C-reducibility.   If these didn't work it was abandoned and
the unavoidable set U modified appropriately.   This may seem a

M-3

very cumbersome approach - especially since circuits like abcdef
but with up to 14 vertices were involved.   (Appel estimates that
the amount of work goes up by a factor of 4 for each extra vertex
in the circuit.)   It might seem that it is better to test for
C-reducibility first.   But in practice this involves a lot of dup-
lication of effort if the first substitute configuration doesn't
work; and it is quicker to start by listing all the colour schemes
to see which can be made good.

Conclusion

The main point I have not explained is the method by which the discharging algorithm and the unavoidable set were modified every time
a configuration could not quickly be proved reducible.   These modifications relied on a large number of empirical rules which have
still not been given adequate theoretical justification, discovered
in the course of a lengthy process of trial and error lasting over
a year.   By then Appel and Haken had developed such a good feeling
for what was likely to work (even though they couldn't always explain why) that they were able to construct the final unavoidable set
without using the computer at all.   This is the crux of their

achievement. Unavoidable sets had been constructed before, and configurations proved reducible; but no one could complete the monumental task of constructing an unavoidable set consisting entirely of irreducible configurations.

The length of the proof is unfortunate, for two reasons. First, it makes it hard to verify. A long proof may take a long time to check, and be intellectually accessible to only a few people. This is particularly true if a computer is involved. Before the introduction of computers into mathematics, every proof could be checked by anyone possessing the necessary mental apparatus. Now an expensive computer may be needed too. Appel estimates that it would take 300 hours on a big machine to check all the details. Few mathematicians in Britain have access to this much machine time.

The other big disadvantage of a long proof is that it tends not to give much understanding of why the theorem is true. This is exacerbated if the proof involves numerous separate cases, whether it needs a computer or not. Lecturers may tend to give students the impression that proving theorems is the objective of pure mathematics; but I am sure that many of us agree that proofs are only a means to an end - understanding what is going on. Sometimes a proof is so illuminating that one feels immediately that it explains the 'real reason' for the result being true. It may be unreasonable to expect every theorem to have a proof of this sort, but it seems nonetheless to be a goal worth aiming for. So undoubtedly much work will be done in the next few years to shorten Appel and Haken's proof, and possibly find a more illuminating one. (It is doubtful that their method can be shortened enough to avoid massive use of the computer.)

In fact, there remain a number of conjectures that would imply the truth of the 4-colour theorem, but do not follow from it. One of these in particular (Hadwiger's Conjecture) is (in my opinion) most unlikely to be provable by the sort of technique that Appel and Haken have used: possibly a shorter proof of the 4-colour theorem may be found from an attack on Hadwiger's Conjecture. None of this, of course, detracts in any way from Appel and Haken's magnificent achievement.

BIBLIOGRAPHY

A slightly expanded version of this article, with references, appeared in the Bulletin of the IMA 14 (1978) 245-299. It also formed the basis for the article *The Appel-Haken proof of the Four-Colour Theorem* which formed chapter 4 of *Selected Topics in Graph Theory*, ed. L.W.Beineke and R.J.Wilson, Academic Press 1978.

* * * * * * * * *

Early news of Appel and Haken's achievement was greeted by the mathematical community with less than unrestrained enthusiasm. One reason for caution is of course the computer involvement: it is extremely easy to make slips in long and involved programs. Now, 5 years later, no such slips have been found; and the program has been checked by a great many people. The expert view is that, if there are any errors, they do not occur in the computer part of the proof. As for the lack of elegance, MANIFOLD-19 remarked: "maybe most theorems are true for rather arbitrary and complicated reasons. Why not?"

*1970 was a remarkable year. The young Russian mathe-
matician Matijasevič solved Hilbert's 10th Problem on
the decidability of Diophantine Equations (negatively)
and found a polynomial formula for primes. And, in
the same year,* MANIFOLD *tackled a related problem:*

Another Formula not Representing Primes

MATTHEW PORDAGE

Fermat thought that $F_n = 2^{2^n}+1$ is always prime, which ought to be
an object lesson to everyone, seeing that $F_5 = 641.6700417$. Many
ingenious tests have been devised to decide whether such numbers are
prime: thus it is known that F_{73}, which has about 3.10^{21} digits, is
divisible by $5.2^{75}+1$. Many, also, are the ingenious formulae
which have been suggested in the search for large prime numbers.
High in the charts are the *Mersenne Numbers* $M_p = 2^p-1$; the present
world record for the largest known prime is held by the United States (who else) with M_{11213} [Now improved].

Then there are the *repunit* numbers $111...1 = (10^n-1)/9$, which
are prime for n = 2, 19, 23; the smallest undecided value is 47.
There is also

$$10^{2n} - 10^n + 1$$

which is prime for n = 2,4,6,8 but *not* 10 (another object lesson!),
and

$$10^{2^n} + 1$$

which is prime for n = 0,1 and composite for n = 2,3,4,5,6. Nobody knows about n = 7.

I was glancing through a table of prime numbers some time ago,
and I noticed that each of 19, 109, 1009, 10009 is prime. Modest-
ly defining the nth *Pordage number* P(n) to be

$$P(n) = 10^n + 9$$

I immediately met the same fate as Fermat, since $P(5) = 7^2.13.157$.

Empiricism is all very well, but a general theory is more satis-
fying; and with a very small amount of theory we can make the Pordage numbers reveal many (alas! not all) of their secrets. Similar
methods will work for numbers represented by any EPIC (Exponential
Plus Integer Constant) formula A^n+B.

The key is Gauss's *modular arithmetic*. Given integers m,n,r we
say that m *is congruent to* n *modulo* r, written

$$m \equiv n \pmod{r},$$

if m-n is a multiple of r. Such congruences can be added, subtrac-
ted, and multiplied *as if they were equalities* (but care is needed
for division!). For example $17 \equiv 2 \pmod 5$ since 17-2 = 15 = 3.5.
Also $6 \equiv 1 \pmod 5$. Adding we get $23 \equiv 3 \pmod 5$; subtracting we
get $11 \equiv 1 \pmod 5$; multiplying, $102 \equiv 2 \pmod 5$. All of these
are correct. Modular arithmetic begins to show its power with

reasoning like:

$$10 \equiv 1 \quad (\text{mod } 3)$$
so
$$10^2 \equiv 1 \quad (\text{mod } 3)$$
$$10^3 \equiv 1 \quad (\text{mod } 3)$$
$$\cdots$$
$$10^n \equiv 1 \quad (\text{mod } 3).$$

So $10^{17}-1 = 99999999999999999$ is a multiple of 3.   On second thoughts, this example is a bit obvious: less obvious is that $8^{13}-1 = 549755813887$ is a multiple of 7.   And it is less obvious still that $1000000000000000000000000000000000000009$ is a multiple of 47. We'll see in a minute how to prove that (*without* doing the division) using congruences.

For a simple case, we'll work mod 7.   We have

$$10 \equiv 3 \quad (\text{mod } 7)$$
$$10^2 \equiv 2 \quad (\text{mod } 7)$$
$$10^3 \equiv 6 \quad (\text{mod } 7)$$
$$10^4 \equiv 4 \quad (\text{mod } 7)$$
$$10^5 \equiv 5 \quad (\text{mod } 7)$$
$$10^6 \equiv 1 \quad (\text{mod } 7).$$

Hence
$$P(6k+5) = 10^{6k+5}+9 = (10^6)^k.10^5+9 \equiv 1^k.5+9 \equiv 14 \equiv 0$$

(mod 7).   That is, for any integer $k > 0$, $P(6k+5)$ is divisible by 7.   Thus $P(5)$, $P(11)$, $P(17)$, $P(23)$,... are composite.

If we try the same trick (mod 5) we don't get anywhere, since always $P(n) \equiv 4$ (mod 5).   Thus 5 *never* divides a Pordage number. Similarly with $2,3,\ldots,11$.   With 13, similar reasoning shows that $P(6k+5)$ is always a multiple of 13.   It seems to be a coincidence that these are precisely the Pordage numbers that are divisible by 7.   At any rate, when we try 17 it is $P(16k+14)$ that is always a multiple of 17.   For primes < 100 we obtain the following table, where p denotes a prime and n shows the $P(n)$ divisible by p.

p	n	p	n	p	n
7	6k+5	23	22k+7	59	58k+35
13	6k+5	29	28k+12	61	60k+54
17	16k+14	47	46k+13	89	44k+21
19	18k+1	53	13k+11	97	96k+52

No other prime < 100 divides any $P(n)$.

When working these out we generally have to do two things: find r such that $10^r \equiv 1$ (mod p), and then s such that $10^s \equiv -9$ (mod p). Then, for any $k > 0$, $P(rk+s) = (10^r)^k.10^s +9 \equiv 0$ (mod p).

We can *always* find r (unless $p = 2$ or 5); indeed $r = p-1$ will do, by a theorem of Fermat which states:

if a prime p does not divide a then $a^{p-1} \equiv 1$ (mod p).

And for certain p we can find smaller values of r (e.g. $r = 6$ when $p = 13$).   In general r divides $p-1$.

The real crunch comes in finding s.   There seems to be no explicit method apart from trial and error.   If 10 is what is known as a *primitive root* mod p then s exists; but s *can* exist when 10 is not a primitive root - and anyway, there is no explicit way to find those p for which 10 is a primitive root.

There is a sort of criterion, though: s *exists only when p divides some Pordage number.* For $10^s \equiv -9 \pmod{p}$ if and only if p divides $10^s+9 = P(s)$. This implies: *if p is a prime dividing P(s) then it divides P(k(p-1)+s)for all* k > 0. In particular *if a prime divides one Pordage number, it divides infinitely many.*

As an example, P(10008k+4) is always divisible by 10009. In particular, P(100084) is a multiple of 10009. It would be futile to verify this by long division: 'long' would be the word!

At this stage of my researches there was a tantalizing gap at P(6). This is about a million, so lies well within the range of computed tables. Unfortunately I was staying with a lady-friend and had carelessly omitted to bring my extensive collection of factor tables with me! By pure chance, though, I did have

(a) A table of primes up to 55079,

(b) *Barlow's Tables* of squares of integers up to 12500.

I resolved to make do with these. A theorem which seemed apposite was: *a prime of the form* 4k+1 *is a sum of two squares in only one way.* Now $P(6) = 1000^2+3^2$; so if I could find another representation as a sum of two squares, P(6) would be composite! And if not, another theorem says it must be a power of a prime, which in this case I could trade up into a proof that P(6) is prime!

One of the endearing things about mathematicians is the extent to which they will go to avoid doing any real work. I wasn't too happy about all those squares, and resolved to cut the number down drastically. Squares can end only in certain pairs of digits; so sums of them can end ...09 only for certain pairs. The only cases are ...00+...09 and ...25+...84. So I only had to look up squares ending 00, 09, 25, 84. A rapid expedition through the tables (it took about 3 minutes!) produced:

$$235^2+972^2 = 55225+944784 = 1000009 = P(6).$$

Thus P(6) *is composite.*

Next job: find the factors. Yet another theorem says that *if* $N = a^2+b^2 = c^2+d^2$ *then*

$$N = \frac{(ac+bd)(ac-bd)}{(a+d)(a-d)}$$

and after all the factors in the denominator are cancelled, N is expressed as a product of two integers.

So I tried a = 235, b = 972, c = 1000, d = 3, and got

$$N = \frac{237916.233084}{232.238} = \frac{293.58271}{17} = \ldots?$$

which doesn't look too good, since 17 divides neither 293 nor 58271. Something had gone wrong! I rechecked my calculations; I even squared 235 and 972 by hand in case the tables were wrong, but no! Finally in despair I tried dividing 1000009 by 293; it went exactly and gave 3413. So now I knew the factors, although something was rather fishy about the method!

Working backwards: 3413.17.4 = 232084 - and that was my error. ac-bd = 235000-2916 = 232084, *not* 233084. The calculation now proceeded:

$$N = \frac{237916.232084}{232.238}$$

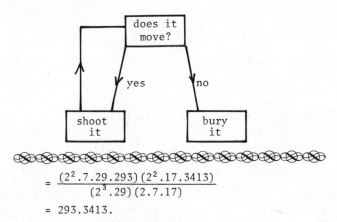

$$= \frac{(2^2.7.29.293)(2^2.17.3413)}{(2^3.29)(2.7.17)}$$

$$= 293.3413.$$

This is the complete factorization of P(6).
   Anyone like to try P(8)?

BIBLIOGRAPHY

Alfred H. Beiler   *Recreations in the Theory of Numbers*, Dover.
L.J.Comrie   *Barlow's Tables*, Spon.
D.N.Lehmer   *Factor Tables for the first ten million*, Hafner.

* * * * * * * * *

*In M-8 Rear Admiral Sir Charles Darlington pointed out some errors
in Matthew Pordage's tables (corrected above) and claimed that P(8)
was prime.   A second letter from him corrected* this: *he showed
that P(8) = 149.671141 is the complete factorization.*
   *Hilbert's 10th Problem, alluded to above, was tackled in M-8 and
again in M-13.   Space forbade its inclusion; but two delightful references are*
J.P.Jones *et al.*   *Diophantine representation of the set of prime
numbers*, American Mathematical Monthly 83 (1976) 449-464.
M.Davis *et al.*   *Hilbert's Tenth Problem*, in *Proceedings of Sympos-
ia in Pure mathematics* 28, American Math. Soc. 1976, 323-378.
*A crucial ingredient of Matijasevič's solution was the Fibonacci
series - yet another triumph for the good-natured rabbit-breeder!*
   *P(9) is still going begging; but the exercise is of questionable
value...*

* * * * * * * * *

M-16

The item of Braziliana on the
right is a perfect example of
rotation, reflection, *and*
translation combined.

SIGNS

"I dreamt the year was 1984, and I was passing a
large comprehensive school.   Children were singing
in the playground:
'Ringrose, Ringrose, Ringrose.
Pascal, Krull, Apostol.
Atiyah! Atiyah! Ahlfors, Dold.'"
So said SIGNS *in* MANIFOLD-2.
It never got any better.

come back, McGonagall, all is forgiven ...

12
10
9
I like twelve
It's got more factors
I like ten
It's easy
I like nine
It's round and square
At the same time.
(Ann M.Atkin.)

Mirror, mirror on the wall,
Who is most symmetrical?

Solomon Grundy
Conjectured on Monday
Hypothesized Tuesday
Existed on Wednesday
Constructed on Thursday
Uniquely on Friday
Contradicted on Saturday
Disproved on Sunday
And that was the end
of Solomon Grundy.

The grand old Duke of York,
He had ten thousand men:
He marched them up to the top of the hill
And he marched them down again.
And when they were down, they were up;
And when they were up, they were down.
It all depends on your point of view
If you turn the coordinates round.

Little Jack Horner
Sat in the corner
Trying to work out π.
He said 'It's minus the logarithm
Of minus one to the i.'

I had a nice conjecture
Nothing would it yield
But a minor theorem
About a finite field.
Andre Weil's students
Came along to hear:
Why any of them bothered
I've really no idea.

Twenty grammes of twopenny rice,
A kilogramme of berets.
Isn't the metric system nice?
Pop! go the ferrets.

Oh...
We know it's true for 1
So we'll use induc-ti-on
Roll me over
Lay me down
And prove it for n.

I do not like thee, decibel!
The reason why: I cannot hear.
But this I know, and know full
well -
I do not like thee, decibel!

Twinkle,
Little
Empty set:
Have you
Any
Members
Yet?

Top, Top, Topology!
Holes are found by homology!
Orientations
Cause complications;
So use Z_2 on all occasions.

MANIFOLD *was intrigued not just by mathematics,*   M-12
but by the mentality of mathematicians.   What is
it like to be a research mathematician?   Here's
one answer:

AN ODD EVENING

IAN STEWART

*As dusk settles gently over the undulating English countryside we
find our hero Rosen Crantz, research student, discussing his latest
ideas with his supervisor, Prof. Guilden Stern, a none-too-success-
ful number-theorist.*

Crantz: Guilden, I'm stuck on my research problem.
Stern: What, the one about prime numbers?
Crantz: Yes.   I was going to prove it for each prime number in
           turn, using that paper of Randy and Hartlisnujam...
Stern: You mean *A Complete List of all Prime Numbers*, Journal of
           Infinity, volumes 173 onwards?
Crantz: Yes, but they've only published the *even* primes so far - I
           think they got stuck somewhere.
Stern: I had a letter from Hartlisnujam a few weeks ago.   He
           said they'd started off well with 2 - that's prime, of
           course - and they decided to run through all the even num-
           bers first in hope of finding some more.   He said they'd
           got up to about 1355579014264890988 but hadn't found any.
Crantz: Perhaps there aren't any other even primes.
Stern: But what about that theorem of Dirichlet's - you know, the
           one that says there are an infinite number of primes in
           any arithmetical progression.   The even numbers form an
           arithmetical progression, don't they?
Crantz: I guess so.   I've forgotten most of what I did at school.
           It's very puzzling.
Stern: Perhaps Dirichlet made a mistake?   He did with his prin-
           ciple, you know.
Crantz: Wasn't that Riemann?   Anyhow, it seems unlikely.   Maybe
           we could *prove* there exist infinitely many even primes?
Stern: By modifying Euclid's proof for arbitrary primes, you
           mean?
Crantz: Exactly.   We'll work with just *even* primes and see what
           happens.   Suppose there's only a finite number...
Stern: We can miss out 2, we know about that...
Crantz: So let's suppose there are only finitely many even primes
           greater than 2, say $p_1,...,p_n$.   Now what?   Euclid forms
           $P = p_1...p_n+1$ and...
Stern: That won't work: it's *odd*.
Crantz: *Very* odd.
Stern: Ha.   So why not define $P = p_1...p_n + 2$?
Crantz: OK.   Then P is even so it must be divisible by some even

	prime, say q. And q can't be any of the p's since they leave a remainder 2 when you divide P by them...
Stern:	... and it can't be 2, since if 2 divides P then it divides $p_1...p_n$ as well, so it divides one of the p's... but that p is prime and greater than 2 so it can't be divisible by 2.
Crantz:	So q is an even prime not equal to 2, $p_1,...,p_n$...
Stern:	Contrary to our assumption. So there must be an infinite number of even primes altogether.
Crantz:	I guess that does it. Dirichlet was right after all.
Stern:	I'll write to Hartlisnujam about it.
Crantz:	I wonder if it'll help my problem?
Stern:	What *is* your problem?
Crantz:	Uh... well... I think my girlfriend is...
Stern:	Your *research* problem.
Crantz:	Oh, yeah. It's a sort of converse to Goldbach's Conjecture.
Stern:	You mean "every even number is the sum of two primes"?
Crantz:	Yes. I want to prove that every prime is the sum of two even numbers. You see, if I could prove that, then...
Stern:	But it's false, surely? What about 3? If 3 is a sum of two even numbers, then one of them is 2... so the other is 1. And that's odd.
Crantz:	*Very* odd.
Stern:	Ha. You need extra hypotheses. Why not assume your prime is *even*?
Crantz:	I thought of that. But suppose we take an even prime q and assume that q = x+y where x and y are even - say x=2u and y=2v. Then q = 2(u+v) so 2 divides q. But q is prime - contradiction.
Stern:	So that disproves it for even primes.
Crantz:	Does it? I never realised...
Stern:	Which means you need only look at odd primes.
Crantz:	But I can't wait for Randy and Hartlisnujam to get to *them*...
Stern:	Well, anyway, you've disposed of half the possible cases.
Crantz:	Plus 3, which you did.
Stern:	Then write it up and publish it. That way, if you do work out the odd ones, you get two papers out of it.
Crantz:	I thought they *weighed* publications, rather than counting them?
Stern:	No, that was before they started printing *Mosaic* on stone tablets. No; five papers and you're a lecturer, fifteen a senior lec-
Crantz:	Wait! Wait! Where in the proof have we assumed that q is even?
Stern:	Oh, where we - no. We didn't. We haven't! The same proof works for odd primes too!
Crantz:	I can see it now! *Falsity of the Converse Goldbach Conjecture* by R.Crantz -
Stern:	And G.Stern...
Crantz:	Yes. We could publish it in the *Notices*...
Stern:	The *Journal*...
Crantz:	The *Bulletin*...
Stern:	The *Proceedings*...

Crantz: The *Transactions*...
Stern: The *Annals!*
Crantz: ....*Ivanov Gos. Ped. Inst. Uc. Zap. Fiz.-Mat. Nauki* –
Stern: *(Thumping him on the back)* Nasty cough you've got there.
Crantz: What a reference!
Stern: Fame! Fame at last! Oh, wait till I see Stevie Smale...
Crantz: We can present it at the International Congress of Mathem-
aticians. We might get a Fields Medal.
Stern: *Two* Fields Medals.
Crantz: I'll be a Professor in no time. They make thousands, you
know. Absolutely rolling in it. I hear one of them rec-
ently sold his 13th century cellar...
Stern: No! Really?
Crantz: And I won't even have to write thirty-one papers and two –
Stern: I could do a lecture tour of the USA!
Crantz: A sort of Malcolm Muggeridge?
Stern: Not exactly; more a Charles Dickens or a – what was that
American chap's name?
Crantz: Twain?
Stern: No, I'll hire a car.
Crantz: And I could do a tour of Paris - lunch at the Sorbonne,
dinner at the Institut - I might even get to meet Bourbaki!
Yes! Yes! *(He pauses, suddenly puzzled.)* *Wait* a min-
ute. What about 2?
Stern: 2?
Crantz: 2.
Stern: What of it? Go on, go on!
Crantz: 2 = 0+2.
Stern: Brilliant.
Crantz: 2 is prime. 0 and 2 are even.
Stern: Oh, BOTHER!
Crantz: Maybe we could patch it up...
Stern: But where have we assumed things are non-zero? I don't
see it. It's odd.
Crantz: *Very* odd.

gruppen

M-4

knit yourself a KLEIN BOTTLE

JANIS WANSTALL

Using three size 10 needles, cast on 90 stitches (30 to each needle).

- *Knit straight until work measures 4 inches (10 cm.).*

- *At beginning of next round, knit 90 stitches, turn and purl 90 (so as to leave a hole in the work).*

- *Repeat these two rows until hole measures one and a half inches (4 cm.). Join round for one row.*

- *Decrease 1 stitch at both ends of each needle on every alternate row until 27 stitches remain (9 to each needle).*

- *Knit straight for a further 12 inches.*

- *Pass work through hole.*

- *Increase one stitch at each end of every needle until there are 90 stitches (30 to each needle).*

- *Knit straight for 6 rows.*

- *Using a fourth needle, take one stitch from needle and one from cast-on edge and knit together - repeat for 90 stitches.*

- *Cast off.*

- *This uses approximately 3 oz. (100 gm.) of double-knitting wool.*

* * * * * * * * *

HINT 3

DO NOT READ THIS NOTE UNTIL YOU HAVE READ PAGE 58.

The essence of a problem of this nature is to prove that certain superficially possible moves are in fact illegal, or rather impossible.

As pointers to the satisfactory solution of the problem, we remark that a certain number of pieces have left the board.   Every time a pawn moved sideways, it could only have done so by capturing. Hence the pawn moves and the missing pieces are somehow connected.

The full solution, by Stanley Collings, appears on page 93.

*"The smallest number that cannot be defined in
fewer than fourteen English words" has just been
defined in thirteen.   There's something very
funny about big numbers.   Maybe there aren't any.*

Beyond the Bounds of Possibility

MICHAEL FORRESTER

Kronecker maintained that 'God made the integers, all else is the
work of Man'.   This epigram expressed his belief that all of mathematics should be based on properties of the integers, a system
which he felt to be free of contradictions.   He was particularly
suspicious of Cantor's Set Theory, which he criticized roundly, on
the philosophical grounds that the notion of a set is far too vague
to be suitable as a mathematical foundation stone.

With the discovery of Russell's Paradox (is the set of all sets
that are not members of themselves a member of itself?) came the end
of the naive period of Set Theory, which led to the introduction of
complicated axiom systems.   The Hilbert programme for proving arithmetic consistent ran into fundamental trouble in the form of the
Gödel Incompleteness Theorem, which implies (a) that if arithmetic
is consistent then there are theorems in it that can neither be proved or disproved, and (b) it can never be proved consistent.

Attempts made so far to circumvent this problem rely on using a
purely constructivist approach to mathematics - saying for example
that a set is not defined unless (a) a rule is given for constructing all its elements, and (b) a procedure is given to decide algo-
rithmically whether a given element belongs to it.

My thesis is that all of these attempts, far from being overcau-
tious, have in fact not been drastic enough; and that instead of a
constructivist approach to arithmetic, it is necessary to adopt a
finitistic theory.   Kronecker believed in the infinitude of the
integers.   I find it hard to see why, once you allow one infinite
set, you should not be able to allow other infinite sets.   The problem is the notion of infinity, rather than any special infinite
set; and Gödel's Theorem demonstrates the problem clearly.   What
makes arithmetic incomplete is simply that statements about *all* integers, if sufficiently complex, can only be proved by an examination of infinitely many cases; and this is impossible.

One of the lessons that physics has painfully learned over the
past few centuries is that our intuition is not to be trusted outside the sphere in which it was formed.   Primitive observations of
stones bouncing off rocks need not apply to electrons bouncing off
nuclei.   Although there is no intuitive barrier preventing arbitrarily large speeds, it appears that in reality the speed of light is
an upper limit.

Another of the lessons that physics is beginning to learn is that
it is best not to put unphysical assumptions into the mathematical

language being used.   Poston [2] has attacked the use of the real number continuum and the idea of 'arbitrary smallness' on just these grounds; if it is not possible to measure a distance smaller than 0.0000000000001 cm., then why use mathematics that says that you can?

In decimal notation the largest number that can be printed on one page of an average-sized book is about

$$10^{2500}$$

so that in 400 pages the largest number we can write in decimal is

$$10^{10^6}.$$

Such a book would occupy a volume of about 20 cubic inches.   A cube of side one light year will contain at most

$$10^{54}$$

such books, so the largest number capable of being written in decimal notation within such a cube is at most

$$10^{10^{10^{60}}} = A, \text{ say.}$$

This may help put in perspective certain results in number theory, such as that of Skewes that if $\pi(x)$ is the number of primes less than x and $li(x)$ is the logarithmic integral function, then $\pi(x)-li(x)$ is positive for some x less than

$$10^{10^{10^{10^{29}}}},$$

a figure he later improved to

$$10^{10^{10^{10^{34}}}}.$$

It also shows the advantage of a compact notation such as repeated exponentials.   However, even using such notation, the best we can do within our light-year cube is

$$10^{10^{10^{10^{10^{\cdot^{\cdot^{\cdot}}}}}}}$$

with A exponentiations, A being as above.   Call this number B. Certainly, by improving our notation, we could write down numbers bigger than B.   But imagine, if you will, all the mathematicians of all time (for the next 100000000000000 years, say) writing out definitions for a highly compact notation for the integers.   By the end of it, they will have written out only finitely many integers; the set of integers they have written out will therefore be bounded above by some integer C.

And yet, almost all integers are bigger than this C.

Now it seems to me that most of the properties we expect from integers are based on ideas got from counting spearheads, wives, or apples or the like.   Commutativity of addition, for example, rests on the fact that if we have m apples and another n apples it doesn't matter which way we put them together.   The idea that n+1 is always bigger than n rests upon the feeling that if we have n apples, we can always add another apple.

This is grand for n = 2,3,4,5,... or so apples; and good old

Hindu-Arabic arithmetic checks it out for numbers of apples up to
millions.   We expect it *always* to be true; and we build it into
our axioms for the integers.   But as I said above, our intuition
is not to be trusted outside the sphere in which it was formed.
And our conceptual Hindu-Arabic arithmetic is not to be trusted
either, given that the whole concept of decimal notation and manipulation rests in the end on intuitions about breaking sets up into smaller sets.   By the time we get up to numbers like A our decimal notation does not apply in any real sense; by the time we
reach B the multiple exponentials have broken down; and by the time
we get to C any conceivable system of notation has broken down.

I have never seen A apples; you have never seen A apples; it is
questionable whether the universe contains A apples - indeed if
Eddington [1] is right it can't even contain

$$10^{10^2}$$

apples, let alone A.   By what superhuman leap of faith do we then
assume that numbers as large as A behave just like small ones?
That they mean anything at all?   What makes us think that A+1 is
bigger than A?   We can't even get A apples, let alone another one
to add on!

Of course, at first sight it is eminently reasonable to assume
that A+1 is greater than A.   We have all the beautiful apparatus
of mathematical logic to prove it.   But our logic is full of nasty
holes.   We only believe it because it supports our naive view of
what numbers are and what they do, based on known properties of
small numbers.   And even if the mathematical system of integers is
OK, there is no guarantee whatever that it applies to the real
world.

What would happen if we based our intuition on the real world,
instead of an airy-fairy version of Pythagorean mysticism?   Let's
take a crate of α apples for some fixed α and do some mental experiments.   Our numbers would be $1,2,3,\ldots, \alpha$ and no more.   Now, in
the mathematical integers, to add two integers we take disjoint
sets with the right cardinalities, form the union, and take the car-
dinality of that.   With apples, we may not be able to make the
sets disjoint (e.g. adding 2 to $\alpha-1$, or 1 to α) so we must do the
best we can and make them as disjoint as possible.   If we denote by
$\oplus$ the resulting 'addition' we find that
$$a \oplus b = a + b \quad \text{if } a + b \leq \alpha$$
$$a \oplus b = \alpha \quad\quad\ \text{if } a + b \geq \alpha.$$

Similarly multiplication becomes
$$a \otimes b = ab \quad \text{if } ab < \alpha$$
$$a \otimes b = \alpha \quad \text{if } ab \geq \alpha$$
where we understand the symbols on the right as ordinary integer arithmetic.

The resulting algebraic system, which we denote by $Z|\alpha$ (the integers Z *truncated* at α) shares many of the usual properties of Z. It is a semiring (in the sense that addition is commutative and associative; so is multiplication; and the distributive laws hold). Indeed, $Z|\alpha$ is the quotient semiring of the positive integers by the ideal of all integers greater than or equal to α. It is therefore mathematically respectable.

Systems like $Z|\alpha$ do crop up. The Hottentots count "one, two, three, many" and effectively work in $Z|4$. With a crate of apples, the arithmetic is $Z|\alpha$. In a universe which contains at most β *things* (and no amount of experiment can ever demonstrate that it contains infinitely many things) the arithmetic will be something like $Z|\beta$. [Other systems occur. My young son was once heard counting one caravan, 'nother caravan, 'nother caravan; *lots* of caravans - working in something like a one-point compactification of a non-Hausdorff version of $Z|2$, or maybe $Z|3$.]

Nor do we lose much by working in $Z|\alpha$. If $\alpha > 3$, it is still true that $2 \oplus 2 = 4$. Prime factorization and uniqueness thereof holds for all numbers *except* α. Subtraction is possible to an extent limited by α; if we make α very large, none of our cherished beliefs will be disturbed except the one about n+1 being bigger that n; and even this only goes wrong for n = α. By making α sufficiently large that we can *never write it down* (e.g. α = C) we obtain a finite arithmetical system agreeing experimentally with all the evidence that can ever be accumulated about integers in the real world. But as a mathematical system it has the advantage of finiteness, and its consistency is not in doubt (to a mathematician), as is that of Z. (However, in a mathematics based on $Z|\alpha$, the problem might arise once more...)

It will of course be necessary to define $Z|\alpha$ without recourse to Z, but this can be done. We shall have to re-orientate the whole of mathematics to accommodate $Z|\alpha$, find replacements for the rationals and reals... but if we wish to work with a mathematics based on sound intuition the step must be taken. I look forward to the brave new future when all this infinitistic nonsense is thrown overboard once and for all.

BIBLIOGRAPHY

1. E.Kasner and J.Newman *Mathematics and the Imagination*. Eddington's theory is discussed on page 32. The whole of Chapter 2 is a discussion of infinity from a viewpoint not unlike that of the present author.
2. T.Poston *Fuzzy Geometry* MANIFOLD-10 1971.
3. This article is dedicated to Errett Bishop.

* * * * * * * * *

There is a superstition on the island of Corfu that if you see a praying mantis, it brings either good luck - or bad luck... ...depending on what happens.

M-7

Lord Russell
shaves all
who don't
shave
themselves.

*"Hit is opunly puplysschid", say the Paston Letters
of 1452. But sometimes the way hit gets puplysschid
is not quite so opun - as a distinguished but semi-
anonymous topologist describes in:*

the Publication System:
a Jaundiced View

dedicated to KATY ADAMS and ANDREW MAY

This is the paper X wrote.

This is the editor, all distraught
Who tore his hair at the horrible thought
Of printing the paper X wrote.

This is the friend whose help was sought
By E, the editor all distraught
Who tore his hair and groaned at the thought
Of the horrible paper X wrote.

This is the proof, all shiny and new,
Of 2.1 and 2.2
Conceived by F, whose help was sought
By E, the editor all distraught
Who tore his hair and groaned at the thought
Of the terrible paper X wrote.

This is the Referee's Report
Which says SUCH THINGS ARE BETTER SHORT
And gives the proof, all shiny and new,
Of 2.1 and 2.2
Proposed by F, whose help was sought
By E, the editor all distraught
Who tore his hair and groaned at the thought
Of the odious paper X wrote.

This Covering Note pretends to be
Detached about the referee.
("He doesn't tell you how to fix
The proof of Theorem 2.6.")

    It quotes the Referee's Report
    Which says Such Things Are Better Short
    And gives the proof, all shiny and new
    Of 2.1 and 2.2
    Proposed by F, whose help was sought
    By E, the editor all distraught
    Who tore his hair and groaned at the thought
    Of the pitiful paper X wrote.

"But still we're sure it can be mended;
If wholly changed it could be splendid."
Typing on a new machine
F answers for *his* magazine.

Signed $E_1 = F_2 = X_3$.

Daisy, Daisy, give me your answer, do:
I'm half crazy, all for the love of you.
It won't be a stylish marriage:
I can't afford a licence...

THINK 13: solutions

1. No. It would be unstable, because a small upward displacement would cause it to clamp to the ceiling, and a small downward displacement would cause it to lose suction and fall.

2. The flow of water is reduced, giving less frictional loss of pressure in the pipe. More pressure is available to accelerate the water.

3. There is a critical minimum size of fire in which the rate of heat release by combustion equals the rate of heat loss to the surroundings. Similarly, small animals have to eat more to keep warm. An atomic reactor is also critical only if it is large enough for the neutron release to balance neutron loss.

4. High bending moments occur in the middle because of dynamic loads and the axial thrust (which enables the brickwork to withstand bending) is reduced. The combined effect produces tensions in the brickwork which it cannot support.

5. American labour charges are higher, and the economies of design dictate shorter journey times at the expense of bigger engines in American vessels. For the same reason, filter-tipped cigarettes are cheaper in this country than non-tipped ones, while the reverse is true on the continent.

6. A bent beam fails due to buckling on the side which is in compression. Buckling is easier to produce in the unsupported edges than in the curved back.

7. Probably not. There would be no difficulty in achieving the correct intertial and elastic properties and outside shape; but considerable difficulty would be encountered in reproducing the correct damping properties. *(Denis Lillee, are you listening?)*

8. The action of wind on sails gives rise to a force on the yacht approximately perpendicular to the wind direction. By sailing at a small angle to the wind, this force has a component in the direction of motion. But the process depends upon balance of the larger lateral component by a force on the keel. Radiation pressure could not be used because there could be no equivalent to the yacht keel in space.

9. Chiefly because the shape is easy to build (it is composed of straight lines) and make strong.

10. The in-out motion required to open or close a hinged door is physically less exacting than lateral movement (generally involving abdominal muscles) as required by a sliding door. But it is doubtful if this is the main reason. It is more likely that hinges are easier and cheaper to make than sliding door gear that dissipates the same energy. Which sort of door involves more kinetic energy when opened in a given time?

11. If a nut is overtightened the threads can be stripped or the bolt-shank sheared. This occurs at a torque that increases with nut size. Assuming the fingers exert a constant force, some safeguard is established by making spanner-length proportional to nut size.

12. No one is quite sure! If relatively little fuel has been used, it may enclose a bubble of gas. The free bubble will probably be an unstable state and will attach itself to the fuel tank wall. The shape of the interface will be such that the sum of the curvatures in two perpendicular directions at all points is a constant, and the bubble is tangential to the wall at points of contact.

13. Some advantages of the ternary system are that the leading digit indicates the sign, so a special symbol for the sign is not required; that correct rounding off is achieved simply by truncating; that to represent numbers of given size, the measure "(number of digits / sign digits)×(possible values of digits)" is less than for any other number base; and that addition and multiplication tables are trivially simple. But ternary digital computers are not used, chiefly because it is easier to build 2-state elements than 3-state ones.

* * * * * * * * *

M-19

one~ move mate : solution

STANLEY COLLINGS

Holmes got straight down to the solution.
   "Only one Black pawn is missing, and this became the Bishop on
e5.   I shall show that on its promotion route it checked the White
King, thereby causing it to move; thus 'White castles' is not a
legal move.   More strictly, I shall suppose that the pawn avoided
giving check, and then demonstrate that the board position is impossible on this hypothesis.
   "The natural and obvious candidate for promotion is Black's QP.
To avoid giving check, this must have promoted on c1 (having captured from b2) or on a1.   In either case 3 pawn captures are required.   In addition, the Bishop originally on f1 must have been
captured by a Knight, and this accounts for all 4 captures by Black.
   "The obvious candidate for the White promotion is the QKtP, and
this must have captured three times to bypass the Black pawns and
reach a White promotion square.   Thus the tally of White captures
is:

by QKtP	3	
by P (e6)	1	creating the doubled pawns
by P (f5)	2	creating the pawn inversion
by Kt (say)	1	capturing B (f8)
	7	

But this is impossible as White only made 6 captures in all."
   "That completes it,"   I exclaimed.   "So Black's QP must have
checked the White King, and castling is illegal.   The solution
must therefore be P x P e.p.   A wonderful exhibition of your analytical skills!"
   Holmes was displeased rather than flattered.   "When have you
known me satisfied by the superficially obvious?   Besides, if you
had more of the analytical prowess you keep extolling in me, you
would not have overlooked an essential part of the data.   White's
QRP moved twice in the course of the last ten moves."   Holmes con-
tinued with his exposition.
   "It is natural to suppose that Black's QP promoted, but it is
not certain that it did so.   It could, *prima facie*, have captured
its way to b5, leaving the QKtP to promote on a1.   If the KtP captured away from the b file before the QP captured on to it, White's
QKtP would have had a clear run to b7, and the earlier deductions
would not hold.   The captured White pieces consist of the Bishop
on f1 (taken by a Knight), a black-square Bishop which could not
have been taken by the supposed QP at b5, and two Knights.

"Therefore the QP took both Knights and the QKtP took White's black-square Bishop... but where?   It must have been after White's QRP reached a4 but before White's QKtP reached b4.   Furthermore, P(b5) must have captured from c6 after White's QKtP got past.

"Counting 10 White moves back from the board position gives:

1. P(a5) - a4
2. B(d5) - a8 = P
3. P(a8) - b7   uncapturing a rook.   Black P(b5)
                                uncapturing Kt.
4. P(b7) - b6   Meanwhile B(e5) unpromotes
5. Kt(b5) clears out of way on a1 to a pawn which ret-
6. Q(b4) clears out of way   racts to a3.   The Black
7. P(b6) - b5   King has to move aside to
8. P(b5) - b4   let this happen.
9. P(b4)- b3   Black P(a3) - b4 uncapturing B.
10. B(a3) clears out of way
11. P(a4) - a3.

Thus QRP can have moved twice in the last 11 moves, but not in the last 10.   Hence it *was* Black's QP which promoted... and so the White King has moved."

Holmes sat back in his chair, but was still brooding.   "The position is still interesting.   If we make the simplest of changes, moving P(c7) to d7, White again has a mate in 1; but now the outcome is different.   I wonder whether that fellow Collings is aware of this final twist....?"

* * * * * * * *

ruler and compass construction

JERRY CORNELIUS

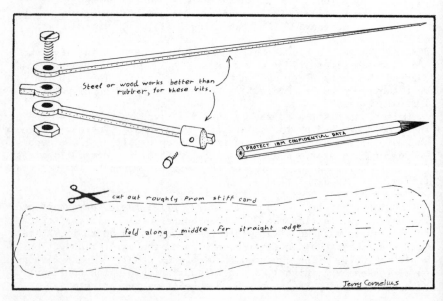

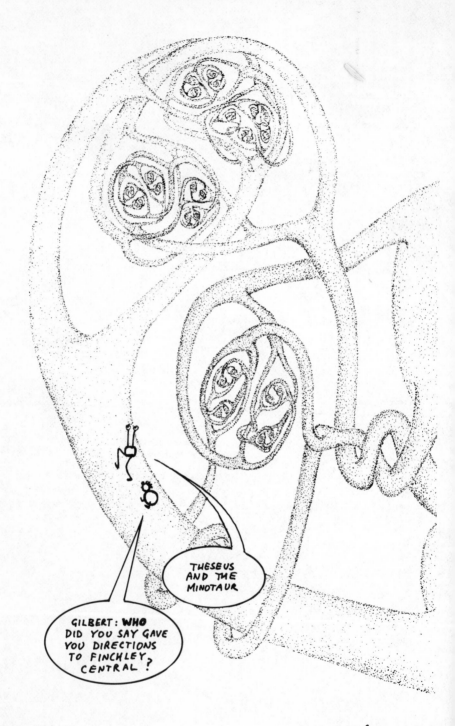